THE INTERNATIONAL TRANSFER OF TECHNOLOGY

Ballinger Series in
BUSINESS IN A GLOBAL ENVIRONMENT

S. Prakash Sethi, Series Editor

Center for Management
Baruch College
The City University of New York

THE INTERNATIONAL TRANSFER OF TECHNOLOGY
Theory, Issues, and Practice

RICHARD D. ROBINSON

BALLINGER PUBLISHING COMPANY
Cambridge, Massachusetts
A Subsidiary of Harper & Row, Publishers, Inc.

Copyright © 1988 by Ballinger Publishing Company. All rights reserved. No part of this publication may be reproduced, stored in a retrieval system, or transmitted in any form or by any means, electronic, mechanical, photocopy, recording or otherwise, without the prior written consent of the publisher.

International Standard Book Number: 0-88730-139-8

Library of Congress Catalog Card Number: 88-11971

Printed in the United States of America

Library of Congress Cataloging-in-Publication Data

Robinson, Richard D., 1921-
    The international transfer of technology : theory, issues, and practice / Richard D. Robinson.
      p. cm. — (Ballinger series in business in a global environment)
    Includes index.
    ISBN 0-88730-139-8
    1. Technology transfer—Economic aspects. I. Title. II. Series.
HC79.T4R59 1988
338.9'26—dc19
                                                    88-11971
                                                          CIP

To the memory of Dr. K. Nagaraja Rao, respected friend and colleague at the Sloan School of Management, who inspired the writing of this volume. Initially, it was to be coauthored, but tragically Raj did not live to participate. Hence, any deficiencies are entirely the author's responsibility. Thank you, Raj, for sharing your insight and wisdom over the years.

Contents

List of Figures   xi

List of Tables   xiii

Introduction   xv

PART I   DEFINITIONAL AND MEASUREMENT PROBLEMS

Chapter 1
Definitions and Measures
of Technology Transfer   3

Technology Transfer Packages   3
Transfer Costs   6

Chapter 2
Dimensions of Technology   11

Dimensions of Technology   11
Transfer Options   17

PART II   THE SUPPLY SIDE

Chapter 3
Sources and Direction of Flow — 23

Sources — 23
Direction of Flow — 24
The Brain Drain Issue — 26
Slowdown in the Commercialization of Technological Innovation? — 27
Technology Flow from the Less Developed Countries — 32

Chapter 4
The Transfer Process — 37

Internal and External Transfer — 38
Value-Added Chain Analysis — 48

Chapter 5
A Summary of Relationships on the Supply Side of International Technology Transfer — 53

PART III   THE DEMAND SIDE

Chapter 6
The Recipients — 61

External Impacts of Technology Transfer — 63
Appropriate Technology — 66
The Cost of Receiving Technology — 67
Risks and Benefits — 77
Propensity to Seek Foreign Technology — 78

PART IV   SPECIAL ISSUES

Chapter 7
Responding to Intervention by Governments and Multigovernment Organizations — 83

Some Examples — 83
Generalizations — 91

Chapter 8
The Siting of Research and Development Abroad — 107

Need, Practice, Justification — 107
R&D Cost — 123
R&D in Less Developed Countries — 124
Host-Government Policy — 125
The Relevance of Corporate Structure — 126

Chapter 9
Protecting Internationally Transferred Technology — 131

Patents — 132
Trademarks — 135
Copyrights — 136
Trade Secrets — 137
International Protection Systems — 138
Trade in Counterfeit or Pirated Goods — 142
Technology Protection in Centrally Planned Economies — 144
The Problem of Technology Protection in Less Developed Countries — 145
Management Problems — 148

Chapter 10
International Competitive Bidding — 151

Alternative Methods — 153
The Bidding Process — 153
Negotiation — 156
The Pricing Problem — 157

Chapter 11
Selecting the Transfer Mode — 165

Types of Joint Enterprises — 167
Partnerships and Strategic Alliances — 170
Strategic Alliances — 174

Chapter 12
Relating Control and Success in Technology Transfer 183

Control and Success 183
Technology Transfer, Structure, and Control 209

A Last Word 213

Index 219

About the Author 227

List of Figures

1–1	The Technology Transfer Package	4
1–2	The Bargaining Window	9
2–1	Technological Dimensions	16
2–2	Degree of Primacy and Completeness of Technology	18
3–1	Advanced Technologies in 2010 Projected by 2,000 Experts	30
4–1	Possible Relationship Between the Propensity of a Firm to Transfer Technology Externally and the Level of R&D	43
4–2	Possible Relationship Between the Propensity of a Firm to Transfer Technology Externally and International Experience	46
4–3	The Value-Added Chain	49
5–1	Relationship of Factors on the Supply Side of the Technology Transfer Process	54
6–1	Relationship of Factors on the Demand Side of the Technology Transfer Process	79
7–1	The Subsidy Continuum	93
7–2	The Project Life Cycle	95
8–1	Civilian Research and Development Expenditures, 1960–1984	108
8–2	Important Criteria for Considering or Not Considering Overseas R&D Locations	118

11-1	Distinctions Between Equity Joint Ventures, Contractual Joint Ventures, and Partnerships	168
11-2	International Partnership or Strategic Alliance	171
11-3	European Economic Interest Grouping (EEIG), July 1985	174
11-4	NADGE Project, Early Warning, Tracking and Interceptor System	179
12-1	Differentiating Joint-Enterprise Type	185
12-2	Selecting the Joint-Enterprise Type	186
12-3	Possible Goal Incongruities	193
12-4	Areas of Control and Some of the Relevant Mechanisms	196
12-5	Some Control Mechanisms	198
12-6	Costs Involved in Participating in a Joint Enterprise	200
12-7	Possible Benefits Derived from Participating in a Joint Enterprise	201

List of Tables

4–1	Mean Returns and Cost of 102 Licensing Agreements	51
7–1	Appendix: Financial Consequences of Tax Incentives and Subsidies When the Parent Government Taxes Worldwide Income	104
8–1	Mean Ratio of the Cost of R&D Units in Europe, Japan, and Canada to That in the United States	124

Introduction

This work is designed to meet the needs of scholars and students, practitioners, and policymakers. For scholars and students, this volume presents a general framework for the study of international technology transfer and, in so doing, pulls together much of the existing insight and research that is scattered over a wide range of publications—managerial, legal, economic, and political. The companion casebook (available directly from the author) extends the consideration of issues by providing specific examples of relevant problems.

Given the apparent fact that the practitioner has no compilation of theory, analysis of major issues, nor description of current practice at his or her disposal, this volume is likewise written to satisfy that need. Although there is no desire to include an in-depth discussion of all possible subjects related to the international transfer of technology, perhaps enough is included to alert the practitioner to major problem areas. More specialized works may then be consulted. Certainly there is no intent to give definitive answers in an encyclopaedic way to problems, only to sensitize the reader to the existence of such problems and some possible ways for their resolution.

Finally, anyone involved in the public policy area relating at all to the international transfer of technology may find this book of value. One could make a strong argument that much of the apparent turmoil in international political-economic relations stems from the explosive growth of international flows of technology, be it transmitted by individuals in face-to-face communication, electronic data transmission, the public media, or via the shift in the locale of productive functions—from research and development (R&D) to manufacture. In the transfer process, a wide range of institutions are involved—from governments, universities, non-profit organizations to state and private enterprises. Political interests at both ends of the transfer process may be intimately engaged as governments move to protect what they perceive to be national interests.

The text opens with a discussion of definitional and measurement problems, then moves into the development of a general model of the transfer process, both on the supply side and demand side. There is an attempt to build into that model an explication of the various pressures at work which tend to initiate (or block) technology transfer, whether internal or external to the firm, and to induce the use of one or more of several possible transfer channels. Many influences are at work, some of which are obvious, others which are quite subtle.

The book then moves into an analysis of several relevant issues in somewhat greater depth, such issues as dealing with various forms of government intervention, the siting of R&D, protecting transferred technology from unauthorized use, international competitive bidding, and the selection of the appropriate organizational mode for effecting a transfer. Under this last heading, four organizational modes are distinguished other than the wholly owned or majority-owned subsidiary—the equity-based joint venture, the contractually based joint venture, the partnership, and technical collaboration by contract.

In the companion volume will be found a series of relevant case studies, all but one never before published in a book format. These cases focus on such diverse subjects as the international technical collaboration agreement, the international patent problem, inter-firm negotiation of technology transfer, method of payment, transferring technology to non-market countries, inter-firm collaboration, technology transfer to a less developed country, marketing new product technology in a less developed country, providing venture capital for the transfer process, and protecting one's trademark. Suggested discussion questions are appended to each case. These cases should be as germane to an executive training course as to an academic program.

My overall purpose in presenting this material is to facilitate the search for ways of reducing the cost of international technology transfer, but in a way which stimulates the international traffic in technology and enhances global development of such a nature as to benefit all. Let the reader be the judge of the success of that enterprise.

—Richard D. Robinson
Gig Harbor, Washington

PART I

DEFINITIONAL AND MEASUREMENT PROBLEMS

CHAPTER 1

Definitions and Measures of Technology Transfer

Anyone teaching or doing research relative to the international transfer of technology has, one suspects, felt very uneasy at moments because of the perceived "mushiness" of the subject. This unease possibly arises because there has been no general model or structure for the field; one merely strung information and insight on an invisible thread and hoped that the thread would continue to hold.

Indeed, one cannot even comfortably define a "bit" (or unit) of technology in an objective, unambiguous manner, as one may in counting bushels of potatoes or even numbers of vehicles, although in the latter case one might have to weight the numbers in some fashion by relative horsepower or possibly number of pounds. But even so, meaningful numbers representing quantities are possible, likewise for value, insofar as market-determined prices can be found. But not so when it comes to "disembodied technology," that which is transferred in the form of human skills and knowledge—either verbally, by demonstration, and/or in writing.

TECHNOLOGY TRANSFER PACKAGES

One is inevitably compelled to think in terms of technology transfer "packages." One scholar wrapped the package as diagramed in Figure 1-1. It is useful for purpose of analysis, however, to break out the

4 / DEFINITIONAL AND MEASUREMENT PROBLEMS

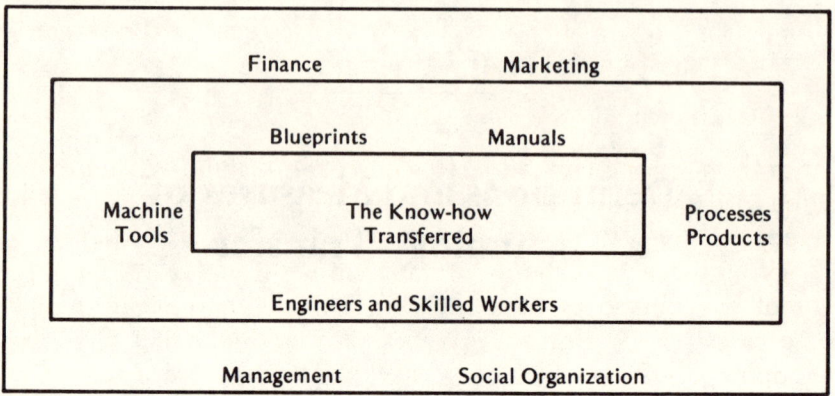

Figure 1-1. The Technology Transfer Package (the Package Within a Package).

Source: Michael Z. Brooke, *Selling Management Services Contracts in International Business* (London: Holt, Rinehart and Winston, 1985), p. 62.

elements within a "technology transfer package" in somewhat greater detail. It may consist of any or all of the following pieces:

1. One or more indivisible technology modules, which may consist of either—

   —"core technology," that which is indispensible to a process or the use of a product or performance of a service, OR

   —"peripheral technology," which is all other modules,

   and which may be transferred via technical documents and blueprints and/or personal explanation, demonstration, or training ("know-how" or technical assistance).

2. Permission to use various rights, knowledge, or assets (that is, under license, franchise, or lease).

3. Hard goods ("embodied" technology), which may take the form of—

   —Capital equipment (although how much of the total cost of such goods represents the technology transfer and how much the cost of the hardware itself may be problematical).

   —Intermediate goods (subject to the same problem), or

   —Final goods (which cause the user to change the way of doing something, from food preparation, to cultivating land, to maintaining an accounting system).

4. Soft goods ("disembodied technology"), which may take the form of—

—written documents, computerized packages, and photographs, or

—oral transmission, whether telephonic, recorded, or face-to-face.

For each component of the package, the technology and/or associated skills may be either *proprietary* (that is, protected by patent, secrecy, trademark, or copyright) or *non-proprietary* (that is, technology legally within the public domain and not the exclusive property of any person or organization). This distinction means that in some cases we are dealing with a legal monopoly that gives rise to monopoly "rents"—which may lead to further distortion of both cost and price. Further, technology may either be leased, with a second party right to use the technology limited in time (also possibly in space—that is, to a specified national market—and to certain specific purposes or uses), or sold outright. In the latter case, the seller has no residual rights to limit application in time, space, or use.

The technology package, or its components, may be transferred either via a direct foreign investment "bundle" or by means of a variety of contractual arrangements, that is "unbundled." Some of these devices are:

1. Export of hardware ("embodied technology").
2. License (for proprietary technology or rights).
3. Technical assistance contract ("know-how" or technical collaboration agreement as narrowly construed).
4. Contract manufacturing with technical assistance, that is, contracting with a second party to manufacture a product in that party's plant.
5. Management contract, which implies a transfer of managerial skill.
6. Marketing agreement, whereby one firm provides marketing services to another, possibly up to and including physical distribution.
7. Training contract (from seminars to inplant or on-the-job).
8. Consulting contract.
9. Architectural and engineering contract ("A&E").
10. Research and development contract ("R&D").
11. Construction supervision contract.

12. Construction contract, which implies the transfer of relevant skills.
13. "Turnkey contract" (construction, plus bringing a plant or project to the point of operation).
14. "Turnkey-plus contract" (a turnkey contract plus the training of local staff to operate and maintain the plant or project).
15. Production-sharing, which involves the export of an industrial plant, for which payment is received in product from the plant.
16. Co-production, which involves the export of some technology, exchange of parts or intermediates in both directions, and production of the final good in both plants—foreign and domestic.

In addition, there is "indirect technology transfer," which occurs when the foreign client either receives technology through the public media, such as professional publications, or comes to the source of the technology, as scholar or student to a university. This indirect transfer may well be a prelude to a formal transfer later on through one or more of the devices listed above. Exposure via indirect transfer may predispose individuals or organizations to seek more formal, direct technology transfer from the same or associated sources—say, from the same country.

The very multiplicity of components in the technology transfer package, and the variety of legal forms which the transfer may take, compounds the problem of developing any precise measure of that which is transferred.

TRANSFER COSTS

It is, of course, possible to speak in terms of the *cost* of a technology transfer and the *price* paid by the recipient, the latter, presumably, being an approximation of the present discounted value of the anticipated stream of royalties or fees and other benefits derived from the technology *as perceived by the recipient* net of all costs. The latter theoretically include all direct costs, as well as appropriate overhead burden and the opportunity cost of foregone profits on exports and direct investment. Obviously, neither the cost nor price can be anything like an exact measure.

Few firms would appear to have a sufficiently refined accounting system in place to enable them to associate their costs with a specific technology transfer, particularly if the transfer is an internal one and not market-mediated. (By "internal" we refer to a transfer from one

DEFINITIONS AND MEASURES OF TECHNOLOGY TRANSFER / 7

part of an organization to another—that is, from one division, branch, or unambiguously controlled subsidiary to another.) Consider the possibly relevant costs:

- An allocation of overhead to cover general promotion expenditures that precede actual negotiation (for example, the buildup of an international image which makes the technology known to, and the acquisition of the technology desirable for, the foreign recipient).
- Allocation of some portion of the underlying research and development (a percentage of total cost, or simply the cost of marginal R&D necessitated by the transfer, such as design modification).
- Negotiating time and associated expenses.
- Possible legal expenses to prevent the illegal use of technology.
- Preparation of technical documents (including possible translation to another language and recalibration into a different measurement system).
- Salaries and transportation cost for personnel responsible for the transfer.
- Cost of interrupting domestic production to train foreign technicians from the enterprise receiving the technology.
- Cost of employing special personnel, such as interpreters, advisors, intermediaries, or consultants.
- Financing costs for the transfer, if any expenses are accumulated before, or in excess of, the stream of payments received. (This includes the cost of any performance bonds or bank guaranties demanded by the recipient.)
- Cost of errors-and-omissions, or non-performance, insurance.
- The value of the market which may be given up by the transferring firm to the recipient of the technology.

Obviously, an exact calculation of the cost of a transfer is difficult. And whether the *price* paid approximates the economic value of a transfer is problematical. Important assumptions must be made, about which reasonable people may well differ.

Payment for many technology transfers rests on a percent-of-sales, or a per-unit-of-production basis. Other transfers become payable upon completion of an estimated percentage of the transfer (for example, in architectural, engineering, construction supervision, construction, and consulting contracts). In any case, future payments must be discounted by both a time and a risk factor, the latter representing the probability of non-payment due either to commercial failure or lack of integrity on the part of the client, or to political intervention by governments. Such intervention may have the effect of either reducing payments (for example, the impact of an unanticipated tax), blocking payment (for example, by denying foreign exchange of value to the supplier), or by prohibiting execution of the contract (for example, politically inspired sanctions of important exports). There is the further risk, of course, that payment will be denied due to the alleged failure of the technology supplier to perform, or of the technology itself to work as promised. Whether *external* arbitration (that is, external to the country of technology recipient) is acceptable and enforceable by the host government must also be factored into the risk calculation by the transferring firm.

So in neither case—the actual cost of the transfer, the present discounted value of anticipated fees, royalties, and other benefits—can calculation be very precise. Frequently, in the transfer of technology to an external (that is, non-related) organization, one speaks of the "bargaining window," the threshold of which is the perceived cost of the transfer by the transferor and the top, the value of the technology as perceived by the recipient. (See Figure 1-2.) Presumably, the latter measure is related to the incremental profit the recipient firm expects to realize by reason of the technology transfer, which is also a very "iffy" sort of number, based on many assumptions about the recipient's costs associated with the transfer (that is, in receiving, adapting, and using the technology), its competitive position (that is, the extent to which it can extract a monopoly rent), the anticipated impact of the new technology on either per-unit cost of production and/or the relative attractiveness of the resulting product to the consumer (which assumes a certain price-demand elasticity).

Therefore, all one can conclude is that the value of that which is transferred is very difficult to calculate with any degree of precision. This fact alone is a major impediment to technology transfer.

The actual form and modality of payment for transferred technology may consist of cost reimbursement, advance payment, lump-sum payment, and/or periodic payment contingent upon some event (or percentage of completion) or sheer passage of time (an annual roy-

Figure 1-2. The Bargaining Window.

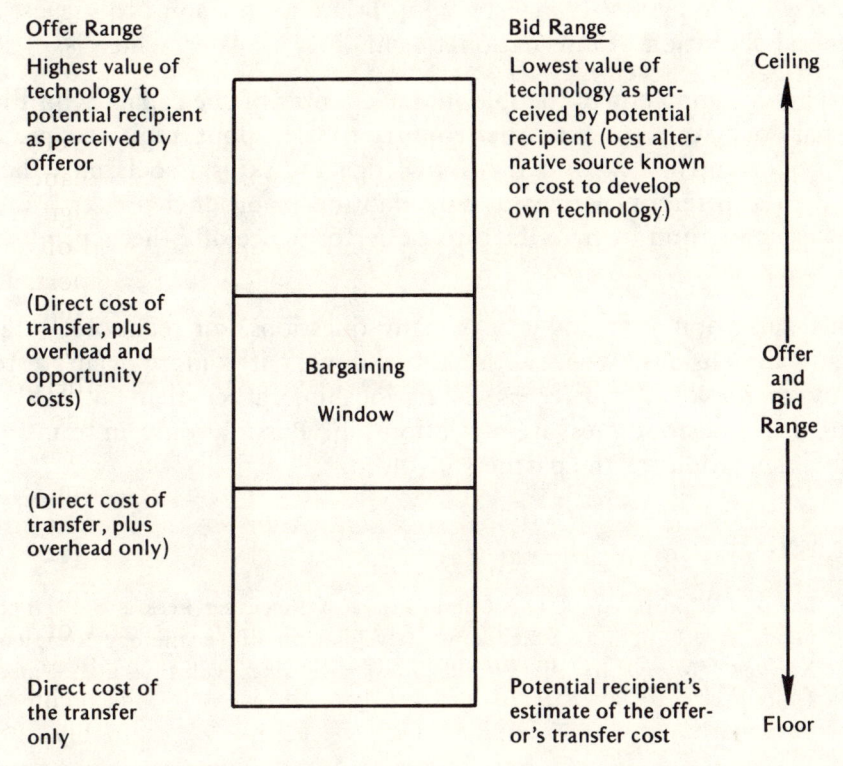

Source: Adapted from Franklin Root, "Some Notes on Price Determination in Licensor/Licensee Negotiations," paper delivered to the Annual Meeting, Academy of International Business, June 1979, mimeo, p. 2.

Note: A major problem lies in the probable fact that the ceiling and floor on the offer and bid sides are not equal. It is very likely that the offeror will value the technology more highly than the bidder (recipient) and that the transfer cost will be seen as higher by the offeror than by the bidder, particularly if the offeror includes any overhead and opportunity cost in the calculation.

alty fee). Payment may be in cash or in goods or services, including a reverse flow of technology (a technology "grant back" or "pooling"). Payment may also take the form of purchase discounts and sales premiums on goods and services bought from and sold to the technology recipient. These counter-flows may coincide with the technology transfer, anticipate it, or follow at some stipulated time. The difficulty of precise measurement, hence, is further compounded.

Despite the difficulty of measuring technology transfer in terms of either quantity or value, one cannot dispute that *something* of commercial value passes. It may be appropriate at this point to suggest a general definition of international technology transfer, which is:

> The development by people in one country of the capacity on the part of nationals of another country to use, adopt, replicate, modify, or further expand the knowledge and skills associated either with a different manner of consumption or product use, or a different method of manufacture or performance of either a product or service.

This definition admittedly begs many questions, for technology has many dimensions. When all is said or done, "it is more accurate to view technology transfer as a relationship, rather than an act."[1] But how does one measure a relationship? Possibly only in terms of the expectations of the partners involved.

NOTE

1. Farok J. Contractor, "The Composition of Licensing Fees and Arrangements as a Function of Economic Development of Technology Recipient Nations," *Journal of International Business Studies*, vol. 11, no. 3, Winter 1980, p. 47.

CHAPTER 2

Dimensions of Technology

Underlying this discussion of the *form or mode* in which technology flows and payment is made is a whole other layer with which to deal. I refer to the inherent differences in the nature of the technology itself, which differences impact heavily on the form and mode of the technical transfer selected and on its price. Indeed, the transfer may, in some cases, be blocked by reason of the nature of the relevant technology. The transfer may be perceived as simply too costly by either the transferring or receiving firm.

DIMENSIONS OF TECHNOLOGY

The first dimension of technology—whether product or process—is its **maturity.** How long has it been around? There is probably a tendency for the more mature to become increasingly labor-intensive. (We define labor-intensity here as the relationship between the amount of labor used per-unit of output compared to the amount of capital used per-unit of output. Capital used may be measured by summing maintenance, obsolescence, depreciation, and interest.) The point is that there is probably a general tendency, as technology-in-use becomes increasingly mature, for the cost of capital equipment to be reduced, its life to be increased, and its maintenance to be less costly. Also, the labor skills required become more widely known and, hence, training and labor cost fall. Labor may then be advantageously substituted for capital, if such is technologically possible and economically attractive.

Still other factors may increase labor intensity of maturing technology, that which has been around for a while and not subject to rapid change. Suppliers of materials, special services, and machines gain experience, and competition among them increases, thereby possibly reducing input costs. The result is lowered working capital requirements. Product markets perhaps become more predictable as the track record lengthens, thereby making possible a better balance between production and sales and, hence, reduced inventories and working capital needs. Investment in relevant R&D may well decline and production time be shortened as experience is gained, both of which changes can reduce capital requirements. Further, the *risk* associated with new production processes, new input requirements, and the development of new markets may be virtually eliminated, and the cost of capital consequentially reduced to the firm. Returns are known and perhaps subject to less variance than in the case of the less mature technology. The net result of all these factors may well be a decline in capital usage and, hence, greater labor intensity. Of course, there can be exceptions; we speak here of the general case.

A second, and closely related dimension is the speed at which technology is expected to change, that is, its **dynamic** quality. Even though a technology-in-use may have been around a long while—and hence, mature—something new may be on the drawing boards waiting on a propitious moment for introduction. Knowledge of expected change may well alter the behavior of both sellers and buyers of present technology. For instance, it can inhibit others from trying to innovate.

A related, but nonetheless distinct, dimension of technology has to do with its **relative importance**. Most important, as the phrase is used here, is *basic technology*, that is, new technology on which basis an entire new industry is created. Chronologically, the next type of technology to appear tends to be *incremental technology*, which may lead in time to *branching technology*. And finally, there may be *major improvement technology*, which still represents a continuity of development from the basic technology but can, nonetheless, be a significant new invention. Examples may clarify.

The first aircraft represented basic technology, as likewise the first use of a vaccine or of an antibiotic. Incremental technology resulted in ever faster, larger, and safer aircraft; in more efficient and safer ways to produce and use the vaccine or antibiotic. Branching technology, in the case of aircraft, came in the form of helicopters and a variety of special purpose craft; in the case of vaccine and antibiotic, the development of a variety of products to treat different diseases. Major improvement technology came in the form of jet-propelled aircraft and synthesized vaccines and antibiotics. One very distinguished observer in listing examples of basic technologies, has written:

> It seems to me that since 1960 there has not been a major, radically new, commercially successful, technological innovation comparable to aircraft, television, nylon, antibiotics, computers, or solid-state electronics; . . . aircraft were not an evolutionary extension of railroads. Television could not be introduced by small improvements to radios.[1]

Another dimension of technology is its degree of **environmental specificity**. Home heating devices are only relevant to cooler climates. Refrigeration is in greatest demand where it is hot. Obvious. Not so obvious perhaps is the relationship between the intensity of use of particular inputs required by a given technology and the relative prices of those inputs in various environments. One might speak of energy-intensive technology, oil-intensive technology, temperature-sensitive technology, water-intensive technology, toxic waste–intensive technology, space-intensive technology, and skill-intensive technology. Cost obviously varies from region to region, from country to country, in respect to the requirements implied in each case.

A fifth dimension is **factor substitutability**—that is, the degree to which factors are substitutable for one another. If the technology involves a chemical reaction in which only specified chemicals are combined in a certain way, then perhaps no substitution is possible, *given the present state of knowledge*. However, many goods may be moved either manually or by machine. Earth may be shovelled or bulldozed. Cigarettes may be hand-rolled or rolled by machines of increasing speed, from low-volume, manually operated machines to high-volume, fully automated machines. The economic choice depends entirely upon relative factor cost, size of the relevant market, and the benefit to be derived from a product standardized to a specific quality within narrow tolerances. Management, however, may find it easier simply to apply a known technology developed perhaps where factor costs differed, or be influenced in its choice by the engineer's propensity to deal with more sophisticated technology regardless of higher factor costs.[2]

Scale specificity is a sixth dimension of technology. Here, one has to use a bit of caution so as to separate scale factors embedded in natural laws and those embedded in the present state of technology. Note the provision in the preceding paragraph, "given the present state of knowledge."

Off-hand, I cannot come up with an example of which I am completely certain of a scale factor determined entirely by natural law. But let us assume that in order to fire a particularly fine porcelain, a temperature of, say, 2700°F were needed. Given the energy to produce that heat (and its cost), and given the market for fine porcelain, we can derive a minimum scale of production which would be economically feasible. (But even in that example, we have a technological parameter that may be subject to change—that is, the cost of producing the energy to generate the 2700°F temperature.) Many of our production scales are a requirement laid down by the state of technological development, not by natural law. There is nothing in nature which says that one can manufacture motor vehicles at minimum unit cost only in runs of 200,000—or whatever the current number is.

A seventh dimension is obvious from the foregoing, the ready **availability** of the desired technology. One may *want* to manufacture vehicles economically in runs of 200, but the technology to do so at a feasible cost is simply not known. This is not to say that an investment in research and development designed to come up with an answer might not generate the appropriate technology, but such an effort obviously adds to cost and time. The effort is thus, very frequently, not undertaken.

An eighth dimension of technology lies in its degree of **complexity**, that is, the relative difficulty of achieving full mastery of the technology, whether in use or design—say, as between a canoe and a hydrofoil. This example does not suggest that the proper handling of a canoe can be achieved without effort and experience, but the *technology* involved is relatively simple.

Next, the technology may be "core" technology in the sense of relating directly to the firm's principal business or, on the other hand, it may be of a by-product nature, or "peripheral," if it does not. We label this factor **centrality** of the technology, which obviously represents a continuum from key central to largely irrelevant peripheral.

Some authors have introduced two other dimensions useful in describing technology, one of which relates to the **continuity of production**. This phrase refers to the fact that technology may be designed for custom or batch production up to and including technology for continuous flow production. In the continuous flow case, the mass production of a highly standardized product may be economically required. Batch production may permit greater flexibility in scale of production and in the design of the final product, thereby making it possible for the producer to respond to smaller and more varied markets. Another dimension of technology is the degree to which a technology is **susceptible to reverse engineering**. Technology that can be analyzed and then synthesized commercially without assistance from those with prior knowledge and/or skill is obviously much more difficult to protect from unauthorized use than technology requiring expensive and time-consuming research and development to duplicate. The latter type is characteristic of process technology requiring a combination of variables (inputs and manipulations) undertaken in a given sequence within a controlled environment, particularly if inputs are used that are not detectable in the final product, as, for example, catalysts.

Hence, we derive a twelfth dimension in describing technology—is it **process or product** technology? This distinction obviously relates to the degree of primacy of a technology, which is described below, but is nonetheless distinct from degree of primacy in that the innovator may have no option as to what is available for transfer. That is, a particular process technology may make possible the more efficient production of a number of mature products that are widely known and used in the market. The only distinctive value possessed by the innovating firm may lie in the process, not in the product. Or, the product in which a new technology is embodied may be the output of a number of well-known and widely used processes. The valuable new technology is carried in the design of the product itself. Consider the difference between a cost-reducing way of molding the hulls for small marine craft versus a new hull design that increases the speed and stability of the craft. (It is of interest to note that the development of process technology tends to follow product technology. Thus, leaders in innovation tend to concentrate on the latter.)

At least one scholar adds a thirteenth dimension, **firm specificity**.[3] He explains, "Firm specific technology is not necessarily associated with any product, but is internal to a firm and based on its entire past experience." The relationship between the degree to which a technology is specific to the firm and the cost of transfer is amplified in these words:

> Conveying firm-specific information to licensees [read contractees or recipients] is a far more tenuous undertaking because it is embodied in personnel and often requires personal interaction with the licensee's staff, training, construction, and start-up management. It involves learning by the licensee of the day-to-day techniques of successful production management.[4]

Putting all of this together (Figure 2-1), we suggest that technology can be described usefully using thirteen dimensions. There may well be others, but these would appear to be the most important.

Figure 2-1. Technological Dimensions.

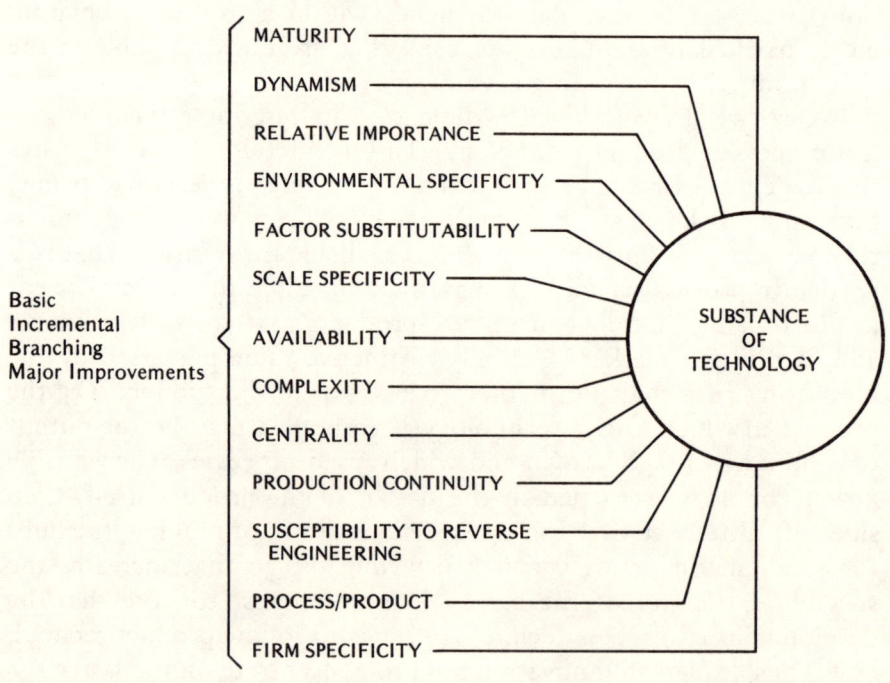

TRANSFER OPTIONS

Closely related to these thirteen dimensions are a pair of transfer policy options that spring from two other characteristics of technology—its **primacy** and **completeness**. Both may be expressed in terms of continua. In either case, the firm may opt to transfer technology represented by points along these continua.

For example, the primacy of technology is defined in a continuum running from *user technology* to *design technology*; completeness, from partial to complete technology. These may be shown graphically, as in Figure 2-2.

How to use a tractor is one thing. How to adapt it to the special conditions under which it is used, another. Still another is the relevant manufacturing technology—from the assembly of knocked down kits to an integrated manufacturing capability. Finally, as the technology moves to increasing primacy, the capacity to modify product and/or process (i.e., design modification technology) is involved, and finally, the capacity to invent (i.e., design technology).

In addition, the transferred technology may be partial or complete, or someplace in between. One may transfer certain elements of user technology relating to the tractor, for example, how to drive and maintain it. But the transfer of *all* of the technology relating to tractor-drawn implements, including data on the various soils, climates, topography, and appropriate crops for each, is another matter. The transferring firm may hold back little or none of the user technology in its effort to build sales. So it is as one moves up the ladder of primacy. The scope of the technology transferred at each degree of primacy can be very limited or virtually complete.

How much of the relevant technology is transferred at any particular level is up to the transferor. The tendency, one suspects, is for the degree of completeness of a transfer to lessen as the technology becomes more primary. (Note dashed line on Figure 2-2.) In this way, the transferring firm maintains leverage against the recipient firm; there is a residual technology retained by the transferor that is desired by the latter—if not now, eventually. At the other end of the scale, toward consumer technology, in most cases there is rarely any reason for holding back any of the relevant technology. One normally wants to create as wide a market for the technology as possible.

18 / DEFINITIONAL AND MEASUREMENT PROBLEMS

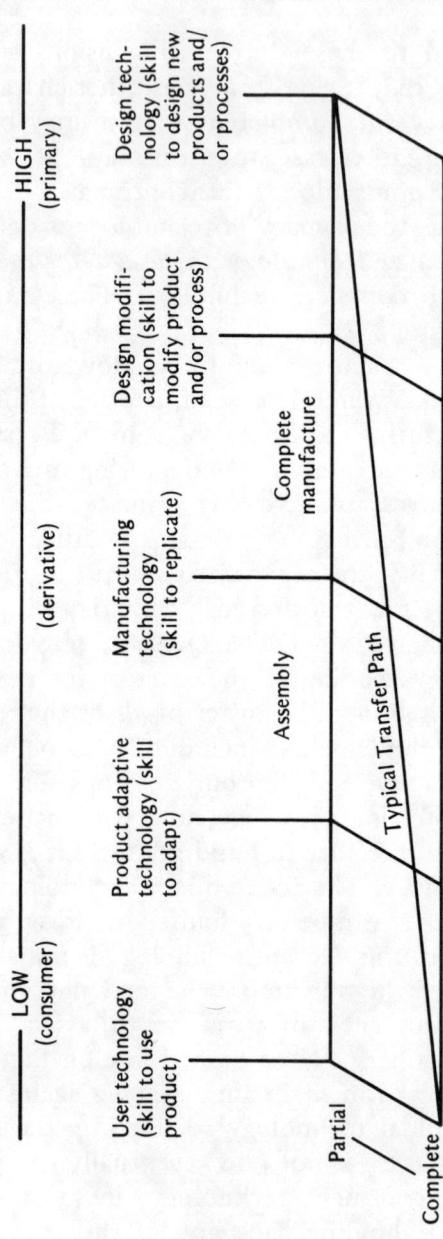

Figure 2-2. Degree of Primacy and Completeness of Technology.

Although these two—primacy and completeness—are, on the one hand, dimensions, they are conceptually different from the thirteen listed previously in that the transferring firm has the option as what to transfer and how far it will go. In the case of the thirteen dimensions, the transferor is given little choice, given the state of known technology, other than the decision as to whether to transfer the technology at all.

Having described, however inadequately, that which flows in the technology transfer process, we turn our attention to the actors (sectors) trafficking in technology across national frontiers—the sources—and to the pressures and restraints which facilitate or inhibit international technology flow. We look first at the supply side. Reference to Figure 5-1 will be useful in the following text.

NOTES

1. Jay W. Forrester, "Changing Economic Patterns," *Technology Review* (Massachusetts Institute of Technology), vol. 80, no. 8, August/September 1978, p. 9.
2. Louis T. Wells, Jr., "Economic Man and Engineering Man," *Public Policy*, vol. 21, Summer 1973, p. 319.
3. Farok J. Contractor, "The Composition of Licensing Fees and Arrangements as a Function of Economic Development of Technology Recipient Nations," *Journal of International Business Studies*, vol. 11, no. 3, Winter 1980, p. 48.
4. *Ibid.*

PART II

THE SUPPLY SIDE

CHAPTER 3

Sources and Direction of Flow

SOURCES

The entities that supply technology (the actors) are essentially four: public international bodies (for example, various agencies of the United Nations), governments (which refers to any government-owned or administered entity), non-profit organizations (NPOs) (that is, foundations, universities, professional and technical associations, charitable agencies), and private business firms. Each is motivated to transfer technology internationally for somewhat different reasons.

International agencies and NPOs are moved by desire to achieve specified objectives (for example, increased literacy, famine relief, medical treatment for the disadvantaged, reduction of infant mortality, reduction in population growth, etc.). *Governments* are motivated by expected political gain (that is, enhanced political power) or by a collective national demand for some sort of action (such as monitoring and reporting on scientific and technological developments abroad).[1] *Private business* is driven by the desire to derive a financial profit. We must be careful, however, in making such distinctions in that large business firms, international agencies, and NPOs are also political organizations and, hence, the power motive may be present to a greater or lesser degree, though presumably not in the primary sense as in ostensibly and exclusively *political* entities.

Having said that about private enterprises, one should also note that state-owned, or parastatal, enterprises are likewise not easy to classify unambiguously as to objective. Some are undoubtedly driven by the need to maximize internal financial results and have no external political objectives. Others may have mixed—possibly conflicting— objectives. For instance, the Dutch State Mining Corporation would appear to be quite dissimilar from the Japan National Railway. The former seems to operate very much like a private profit-maximizing enterprise. The latter appears to be responsive to various political pressures to provide services even though not profitable in an internal financial sense.

Both international agencies, governments and NPOs are often simply the conduits through which technology supplied by business firms flows. In such cases, negotiation takes place between a firm and a public body (international and national) or an NPO, possibly of the same nationality. Indeed, the purchase of technology for export by either national government or NPO may be "tied" in the sense that its source is limited to domestic firms. That is, such grants or loans must be used by the recipient to purchase goods and services from a specified country. Very frequently, firms interested in being considered as potential technology suppliers through non-business intermediaries are encouraged to register with a variety of public agencies and thereby become "qualified" sources. Such registration means that they have demonstrated technical competency and financial integrity to the satisfaction of the registering body and that their bids on advertised projects will be considered.

DIRECTION OF FLOW

It seems quite clear that the bulk of international technology transfer takes place between the more industrialized countries and secondarily, from those countries to the less developed nations. However, scholars have begun to point out an increasingly significant reverse flow, the movement of technology out of the less developed countries (LDCs). Very frequently that flow is into other countries at a similar or lesser level of development, but not exclusively so. Many reasons have been advanced for this increased traffic.[2]

According to the international economic theory, the most capital-intensive inputs to industry should be produced where capital is relatively abundant compared to labor and hence, relatively cheap. In that research and development, and innovation generally, are very capital-intensive activities, one would normally expect that the bulk of commercially valuable innovation would take place in the more industrialized countries which, by definition, are the richest in capital, that is, in accumulated savings invested in high-level skills and capital equipment. In time, the technology so developed would be exported to less developed countries, first embodied in products, then as part of the direct foreign investment package, and finally, as unbundled technology of increasing primacy and completeness transferred under contract.

Research and development is capital-intensive precisely because it is so far removed from, and so tenuously linked to, the production of something of commercial value. If one considers the project life cycle (see Figure 7-2), it is apparent that those activities calling for investment in the early phase of the cycle—(a) opportunity discovery and development, (b) feasibility analysis, (general, marketing, technical, financial), (c) organization of a project team (i.e., recruiting and/ or training persons of requisite skill), (d) technology acquisition (whether purchased from others or as a product of internal research and development), and (e) engineering—are all far removed from commercial production and, therefore, require heavy investment. Furthermore, risk is great at these early stages because of the uncertainty that *anything* of commercial value will flow out of the effort.

Analyzing capital or labor intensity at the individual firm level may be very misleading in that much of the investment in research and development is in the highly skilled persons involved, that is, investment made *by society* collectively, not by a given firm. The social cost of producing a front-line scientist or engineer may run to many hundreds of thousands of dollars. Real resources are used by society in generating such skilled people. These resources take the form of maintenance as well as the application of educational and training resources, which in themselves represent significant investment by a society.

It is precisely the realization of this social investment committed to producing highly skilled people that motivates governments to introduce restrictive emigration policies. It is not only the "iron curtain" countries which have imposed barriers. Such a democratic country as Singapore, for example, tries to restrain the emigration of its talented youth by requiring families of such to post bonds, release from which is only obtained upon the return of the young persons to the country.

THE BRAIN DRAIN ISSUE

Many countries are concerned about the so-called "brain drain," and rightly so. For a relatively poor country to commit scarce resources in developing highly skilled persons, such as scientists or engineers, and then to lose them via emigration to the more industrialized countries because the economic rewards in such countries are greater and living conditions more attractive, represents a very real social loss to the country of origin. The posting of bonds, the holding of families in virtual hostage, the threat of cutting off all financial support from the home country, the sale of exit permits for large sums of money, and the absolute prohibition against emigration for certain categories of people are all devices that countries have employed. A major problem facing the People's Republic of China currently, for example, is how to lure home the talented young men and women it sends overseas for further education and to do so in ways found tolerable by the host countries. These people are precisely those who have mastered skills, have the least ties to the homeland, and who have demonstrated a certain willingness to assume risk, a risk always present when one moves from a familiar cultural environment to the unfamiliar. To try to attract such individuals home by the offer of economic benefits comparable to those they might receive in the richer, more industrialized countries is really not tolerable, because that would place such people in a separate, higher economic class than those who stayed at home. Worse, it is likely to create a cosmopolitan, albeit skilled, elite, members of which would see themselves as inherently more sophisticated—perhaps better in some sense than the stay-at-homes. The loss of those who remain abroad can be made up only if a similar number of equally skilled persons of foreign origin were to immigrate to the less developed countries. Various forms of international technology transfer undertaken by private firms, nonprofit organizations (i.e., foundations, industrial associations, international agencies), and government organizations do sponsor such a reverse movement—from the Ford Foundation, to the Peace Corps, to the specialized agencies of the United Nations, to various agencies of national governments—but the numbers would appear to be very asymmetrical.

Hidden in the aggregate numbers, however, are those individuals carrying technology from the less developed countries to other LDCs and, increasingly, to the more industrialized countries, in commercially viable enterprises. Many arguments have been advanced as to why this is occurring. *The most convincing to me is that, on a worldwide basis, the rate at which the commercialization of really new technology is taking place has slowed.* If so, the average maturity of the technology in use, in a global sense, would be increasing. That would mean that the labor intensity of the technology in use would be increasing, for it can be demonstrated that, on average, the more mature a technology, the greater its labor intensity. That point has already been made in the preceding chapter.

SLOWDOWN IN THE COMMERCIALIZATION OF TECHNOLOGICAL INNOVATION?

Quite apart from the expected impacts of a slowdown in the commercialization of new technology, which will be treated shortly, there are a number of reasons for such to be happening. *First*, one must note the unparalleled investment in fixed industrial assets worldwide. There is little incentive to introduce really new technology until present plant and equipment has paid for itself. The cost of doing otherwise is simply too great. *Second*, the average age of the populations in the more industrialized countries has increased significantly. Some have postulated that such an age increase reduces the demand for much new technology in that investment in consumer durables has already been made. People have grown familiar with present technology and resist learning the skills to utilize the new. *Third*, generally the time-horizon of investors has shortened by reason of their increased age, heightened uncertainty, more frequent and unpredictable government intervention, and fear of inflation. Earnings displaced very far into the future are discounted by large numbers, thereby discouraging R&D where the payoff may be long delayed. *Fourth*, as frontier technology becomes increasingly complex, greater investment is required by society to train frontier technicians, engineers, and scientists. The cost of research and development is pumped up. *Fifth*, there is increasing uncertainty over the outcome of R&D. Will anything of real commercial value, which society will accept, be produced? One has to be concerned with potential toxicity of both product and effluence. *Sixth*, one needs to note that much of the new technology, for many of the reasons noted above, is of a marginal sort—essentially "improvement" technology, whether one speaks of aircraft, vehicles, computers, communications, health-related innovations, or whatever. *Seventh*, there has been a demonstrable breakdown in the patent system worldwide and greater reliance placed on protection by secrecy by innovating individuals and corporations, which means that important inventions or innovations are perhaps disseminated more slowly.

Precisely because of the complexity and marginality of much new technology, plus the sheer volume of technical innovations (even though not of a radically new nature), it has become increasingly difficult to achieve effective protection of an invention via patent. With the apparent slowdown in the commercialization of radically new technology, the worldwide dissemination of relevant technical skills and managerial ability, plus the shift of manufacturing in the direction of some twenty or so newly industrializing countries (NICs), where patents are notoriously ill-protected (if at all), the ease with which many innovations can be copied is greatly facilitated. A patent puts an innovation on public display. The technical description can be read in the relevant patent office, thereby rendering copying—and improvement—easier. Secrecy, on the other hand, keeps the innovation from public view, except insofar as it is subject to reverse engineering. Most industrialized countries, wherein the bulk of commercial innovation occurs, have laws protecting commercially valuable trade secrets, and permit employers to extract pledge-of-secrecy commitment from employees in the form of enforceable contracts.

An *eighth* factor inhibiting the commercialization of really new technology is increasing government regulation—and the rise of legal liability—in relation to product quality, environmental impact, and occupational hazard. And *finally*, there is the fact of labor opposition to the introduction of new technology unless that opposition can be defused by employment commitments and believable promises that the benefits derived from the new technology via cost reduction and/or enhanced sales and profits will be shared equitably with labor.

Let it be noted that many or all of these factors could shift at some point, thereby releasing a whole new generation of radically new, commercially valuable technology on the world. The Systems Dynamics Group at the Massachusetts Institute of Technology maintains that "industrial economies tend to overinvest in existing technologies, especially in capital goods industries, and that excessive investment leads to their decline when demand wanes." They go on to claim that "the latest wave of overinvestment began in the 1960s, but is becoming obvious only now." Another scholar observes that "clusters [of radically new innovations] occur in depressions because investors become willing to invest large amounts of money in high-risk ventures only after the incentive to invest in traditionally profitable ventures disappears... [In a depression] reservations about untried, risky new ideas disappear with the sense that relief might come from anywhere."[3] One attempt to identify the new technologies that could reasonably be expected to make their appearance by the year 2010 is given in Figure 3-1.

If, what we surmise from the preceding arguments is indeed happening—that there is indeed a slowdown in the commercialization of really new technology—then one would expect:

- A gradual loss of competitive advantage by the more industrialized (affluent) countries;
- A shift of manufacturing to the more industrialized, politically stable, less developed countries;
- Erosion of control by multinational corporations and a shift in the direction of more joint venturing, the formation of more partnership and collaboration agreements, and the negotiation of more technology transfer contracting;
- Increasing importance of international service and trading companies;
- Worldwide dissemination of relevant managerial and technical skills;
- A general reduction of investment (as a percentage of total product) in commercially relevant research and development; and
- *The appearance of LDC-based firms active in the international transfer of technology.*

30 / THE SUPPLY SIDE

Figure 3-1. Advanced Technologies in 2010 Projected by 2,000 Experts.

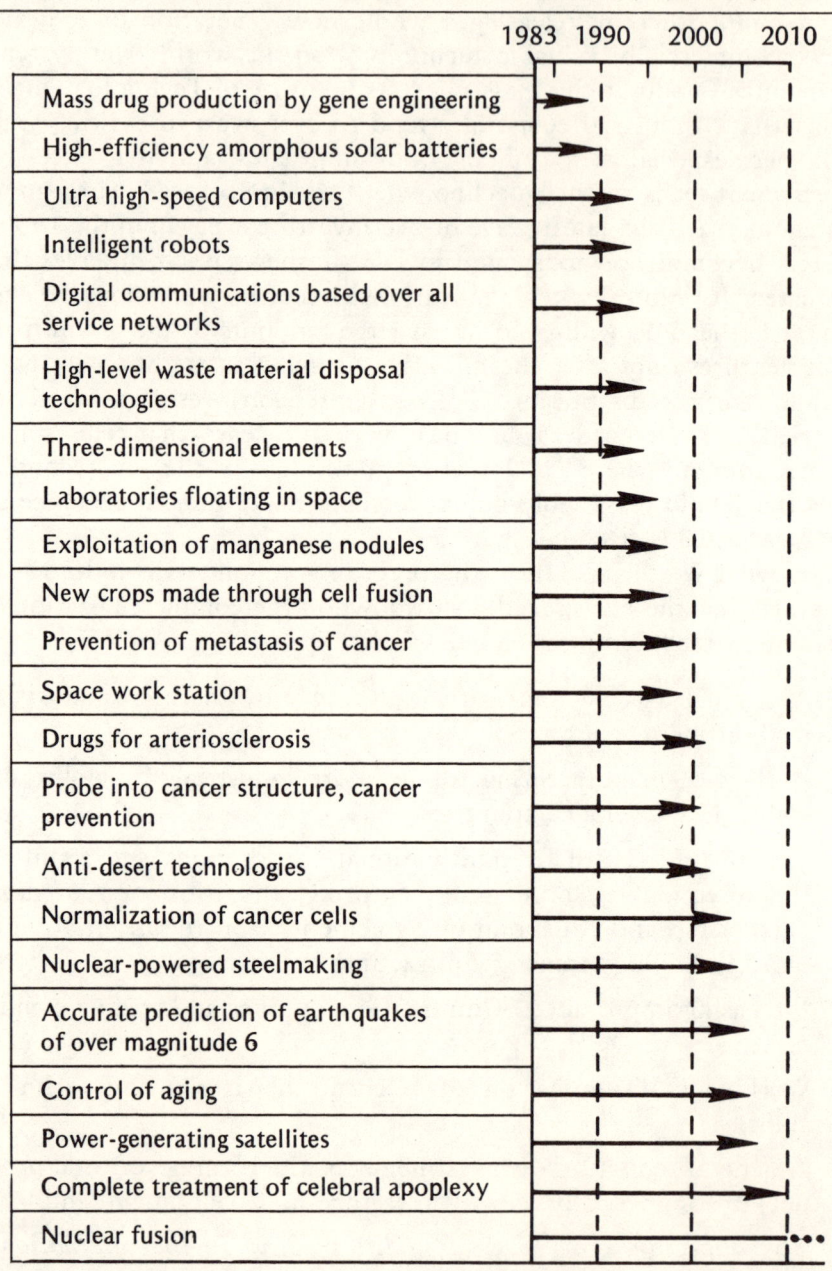

Source: Science and Technology Agency, published in *Japan Economic Journal*, June 25, 1985.

Each of these things seems to be occurring, though the evidence is not yet conclusive. In 1950, something like 96 percent of the world vehicles were produced in the advanced market economies; by 1980, the figure was 82 percent and apparently falling. In 1950, 78 percent of the world's steel production took place in the advanced market economies, by 1980 only 55 percent. In 1963, something like 92 percent of world manufacturing value was created in the more developed countries (market and centrally controlled); in 1980, 89 percent.[4] Perhaps most conclusive is the apparent fact that all of the Organization for Economic Corporation and Development (OECD) countries (i.e., the more industrialized, market economies) taken together reported in 1979 that of all of their imported manufactured goods, 10 percent originated in less developed countries; by 1984, that figure had risen to 14.5 percent (for the U.S., 28.6 percent, Japan 29.4 percent, the European Community, 7.7 percent).[5]

Slower commercialization of technological innovation has been accompanied by a significant shift of the world's manufacturing capacity in the direction of the less developed countries. One would normally expect that new capacity to be concentrated in those LDCs in which there are significant pools of trained people and some degree of political stability—which would encourage long-term investment in fixed industrial assets. Hence, the category of newly industrializing countries (NICs) has emerged in the literature.

The relative importance of foreign direct investment, the hallmark of the multinational corporation, seems to be declining—at least compared to bank lending and other forms of international finance. And there is some evidence, albeit not yet of a compelling nature, that the international joint venture (both equity and contractual), and formation of international partnerships or collaborations, and international technology contracting have increased relative to the establishment of the unambiguously controlled foreign subsidiary, which is the device most often preferred and used by the multinational corporation. Concurrently, there is some indication that international trade in services has been increasing relative to both investment and trade in hard goods. The rapid growth of both the general and specialized trading companies provides some evidence, also the explosive internationalization of financial and communication services (including electronic data flows).

Given the many schools of technology and management which have appeared around the world, it is clear that there has been a rapid dissemination of technical and managerial skills. Also, there is evidence that the number of home-country nationals employed by firms operating internationally has declined steadily in favor of host-country nationals as an increasing number of skilled, local persons have become available at lesser cost. In addition, the cross-border flow of students at all levels has increased dramatically.

There have been numerous studies relating to the percentage of output devoted by countries to commercially relevant research and development. In aggregate, there would appear to be a stagnation, if not a relative decline. That is, less and less of the world's output is devoted to research and development and to training of individuals capable of carrying it on. One has to be cautious with the numbers, for all *militarily* related R&D should be eliminated; its relationship to the creation of *commercially* valuable technology is exceedingly tenuous.

TECHNOLOGY FLOW FROM THE LESS DEVELOPED COUNTRIES

Another expectation, if indeed the commercialization of really new technology has slowed, is that there would be an increasing flow of skills and technology out of certain of the LDCs, specifically the NICs, to other LDCs and increasingly into the more developed countries. This development, it is hypothesized, is due largely to the development of "improvement technology" in these countries as they gain experience and modify technology imported from the more developed countries to their own needs, thereby bringing it closer to the needs of other less developed countries and giving the innovating countries and firms that develop improvement technology an advantage in trade. The result is the rise of the LDC-based multinational corporation.[6] One author identifies such LDC-inspired innovation as "indigenization," by which is meant "making technologies appropriate for the situation where they are to be applied."[7]

Some of the special situations that influence technological innovation are listed in the same study:

- macroeconomic characteristics, such as the economic system (market versus centrally planned economics), factor resource endowment [hence, different prices], stage of economic development, interest rate, inflation rate, unemployment rate, etc.;

- microeconomic characteristics, such as the availability of entrepreneurial talent and capital, supply, demand, and competitive characteristics of industry, acceptable organizational forms, regional and local demand, etc.;
- social and cultural characteristics, such as educational levels, religious beliefs, national aspirations, caste and class structure and traditions; and
- the political environment and the stability of government policies on industrialization.[8]

Various forms of "adaptive technology" necessarily appear over time. They involve adapting existing products (i.e., those imported from the more developed countries) to the needs of the local market, adapting existing process and use technology so as to fit local resource endowments and climatic conditions, and adapting "the technology delivery system and organizational form to the indigenous social, cultural and political environment."[9]

One can also speak usefully of "transformation technology," by which is meant that technological innovation required to transform a *traditional* process into a more productive enterprise. This can be achieved possibly by developing a technology "which increases the overall employment potential of a particular industry but which is capable of being implemented on a scale manageable by local entrepreneurs [and managers]," which increases the use of local materials and skills (which may mean the downgrading of technical complexity and widening tolerances), making production tools and equipment more appropriate, and by "organizing production, marketing and financing in such a way as to use optimally rural labor and marginal farmers on a periodic basis according to when they become available."[10]

Another author lists some of the possible advantages of technology developed by those third-world enterprises that have been able to compete internationally. These technologies are:

1. efficient at smaller scale due to the smaller and usually more concentrated industries in which they operate at home;
2. more labor-intensive due to the low wage relative to capital costs in their home countries;
3. used at higher capacity utilization due to relatively higher capital costs;

4. capable of utilizing more locally produced inputs;
5. more flexible so they can use different inputs and produce several different outputs; and
6. "older" and less R&D-intensive.

As a result, the products of those third-world firms are likely to have lower R&D intensity, lower marketing intensity, lower product quality, and compete on a price basis with lower price-cost *margins*. The overall conclusion is that "since third-world firms compete internationally on a price basis, cost efficiency must be an important component of their strategy. In order to achieve such efficiency, it becomes important that they use their capital equipment at optimal capacity utilization with optimal capital-labor ratios and the optimal type and combinations of inputs to produce the optimal combination of outputs."[11]

This analyst concludes, "These optimization activities imply that third-world TNCs [read, enterprises transferring technology abroad either via direct foreign investment or contracting] need a high level of skilled managers and workers who are experienced in using their technology in order to compete successfully."[12]

The point of this discussion is to underscore the fact that the international traffic in technology is less and less monopolized by the industrialized countries. Furthermore, there is great pressure on firms bringing a new technology to the market to try to penetrate all markets simultaneously in order to recoup R&D investment as speedily as possible. This penetration largely negates the product life-cycle theory, which postulates a sequence of events—innovation, production in the home country (an affluent industrialized society), product export, eventual production abroad via foreign direct investment, disappearance of production in the home country as imports capture the domestic market. Intensified competition to penetrate all potential markets simultaneously strains the capacity of even the largest firms and compels a degree of flexibility of entry strategy, from subsidiary to joint venture, to collaboration or partnership, to contract manufacturing, to international technology transfer contract.

NOTES

1. In early 1983 a fairly vigorous demand was made that the monitoring of Japanese technological innovations be stepped up by the U.S. government.

In 1985, the Japanese-U.S. Conference on High Technology and International Environment was established as a step toward creating a more symmetrical access to technology, that is, greater U.S. access to Japanese technology (*Japan Economic Journal*, November 7, 1987, p. 19).
2. See "Exports of Technology by Newly-Industrialized Countries," a special issue of *World Development*, May/June 1984.
3. "A Technology Lag That Might Stifle Growth," *Business Week*, October 11, 1982, pp. 126–130.
4. Bruce Kogut, "Design Global Strategies," *Sloan Management Review*, Summer 1985, p. 23.
5. *OECD Observer*, March 1986, p. 5.
6. Donald J. Lecraw, "Technology Transfer by Third World TNCs," unpublished paper based on research funded partially by the UN Center on Transnational Corporations, University of Western Ontario, October 1982.
7. Paul Shrivastava, "Technological Innovation in Developing Countries," *Columbia Journal of World Business*, vol. 15, no. 3, Winter 1984, p. 25.
8. *Ibid.*, pp. 25–26.
9. *Ibid.*, p. 26.
10. *Ibid.*, p. 27.
11. Lecraw, *op. cit.*, p. 6.
12. *Ibid.*, pp. 6–7.

CHAPTER 4

The Transfer Process

In that probably most commercially relevant technology moving internationally involves private firms on one or both ends of the transfer, we focus here on the decision-making process within a private firm (or a government-owned enterprise acting as a private firm). The posture of both parent and host governments is one factor in this process. Do the relevant governments actively seek or support the transfer? If so, via an internal or external (i.e., market-mediated) transfer—or are they neutral on that score? And are the governments a factor in the choice of the technology to be transferred? One is aware of such considerations as taxation, rules of competition, export controls, and availability of finance via an economic assistance program of some sort. The host government, on its part, may intervene with taxes and subsidies that penalize or reward various flows, with restraints on foreign ownership and alien employment, with limitations on the remittance of dividends and fees and royalties, and with rules defining which technology may be imported ("appropriate technology"). The various forms of government intervention are dealt with in a later chapter.

In any event, we must deal with three separate but inter-related factors: (1) the propensity of a firm to transfer technology, (2) its choice of the transfer mechanism (i.e., internal or external), and (3) its selection of a particular technology to transfer. All three decisions are probably derivative of the transfer organization's perception of the cost, the risk, and the benefit associated with the transfer.

In that we are addressing ourselves to all forms of international technology transfer, to get into a long analysis of why some firms react positively to foreign market opportunities (indeed, to seek them out actively), and other firms do not, would take us far afield into international trade and investment theory, not to mention organizational theory. It is enough for our purposes here to specify that a firm's propensity to transfer technology internationally is related to its awareness of a foreign demand for its products or services, to its competitive position, to its past history and resources, and to the intervention of parent and host governments. Such intervention can, of course, be either conducive to a technology flow or otherwise. (For more on this, see Chapter 7.)

INTERNAL AND EXTERNAL TRANSFER

The decision of greater significance to us at this point is whether the firm opts to transfer technology *internally* (that is, as part of the direct foreign investment package) or sell or lease the technology *externally* to a completely separate entity, thereby involving an external or market-mediated transfer. To speak of an internal transfer of "bundled technology" probably makes little sense, for in such cases one is referring to the transfer of technology to a firm's own branch or subsidiary (i.e., an unambiguously controlled entity) abroad and, as such, it becomes an integral and undistinguishable part of the firm's investment in that branch or subsidiary. However, for a variety of reasons a firm may nonetheless contract with such branch or subsidiary for the transfer of "unbundled technology," that is, for specific components of the technology package (e.g., use of patents, trademark, know-how, marketing, etc.).

Obviously relevant to the choice between an internal and external transfer is the extent to which—and how—the technology can be protected from unauthorized use. A high degree of protection enables the transferring firm to extract a "rent" from the exploitation of the technology in the foreign market. Can it do so best through a wholly owned or majority-owned subsidiary, a joint venture, a partnership, or via a specific contractual relationship with a second party (that is, a license, technical assistance contract, partnership, marketing agreement, management contract, consulting contract, financial agreement, or what)? The effectiveness of local law regarding patents, trade secrets, copyrights, and trademarks becomes germane, so likewise the perceived enforceability of commercial contracts generally. (See Chapter 9.)

The internal transfer of technology tends to be less costly than a market-mediated transfer for a variety of reasons, some of which are:

1. Persons at both ends of a transfer speak the same language in an organizational and technological sense.
2. Prolonged negotiation as to terms and conditions may be eliminated, likewise most of the associated costs, including legal.
3. The firm runs little risk of non-payment for commercial reasons (that is, lack of integrity or bankruptcy on the part of the recipient).
4. Performance bonds or guarantees are rarely involved.
5. Claims based on failure of the technology to operate as expected are extremely unlikely.

Clearly, in the case of an external transfer, that is, a transfer to an independent firm, these savings are not present unless the firms have a long history of fairly intimate collaboration.

Of considerable interest and perhaps significance on this last score is the recent emergence of a growing number of international partnerships, the so-called strategic alliance. By partnership, I mean an agreement between two or more legal entities—persons or corporations—to work together for a specified purpose under a joint management team. All revenue flows through to the partners after payment of joint management costs. The partnership itself is rarely taxed on its income in that legally it is considered to be simply a conduit. Likewise, all liability flows through to the partners, a liability which may be joint or assigned by agreement either as a percent of the overall liability or by project. A partnership may, if the partners so agree, employ persons (though they may be simply assigned from the partner firms), enter into contracts, borrow, or incur other forms of liability on behalf of the partners. The partnership may be an agent of the partners for specified purposes, but it itself has no earnings, no capital other than that assigned by the partners, and no liability apart from that assumed by the partners. Recurring characteristics of such relationships are (1) little or no joint investment by the partners, (2) some form of joint management of specified functions,

(3) exchange of personnel and intensified inter-firm communication, and (4) long-term commitments. One might well look at these partnerships as attempts by the partners to reduce the cost of external transactions by effectively internalizing them within a long-standing and intimate relationship.

It has been suggested by more than one analyst that organizational linkage between transferor and transferee influences the cost of technology transfer and, hence, presumably, the propensity of firms to become involved. For example, it would appear that transfers to affiliates (i.e., branches and subsidiaries) take place at lower cost than to unrelated entities. "Furthermore, transfers to government enterprises in centrally planned economies involve extra costs because of procedural differences involved and the extra documentation that is generally required."[1]

I would generalize this so as to include all public-sector entities except those operated as pure businesses that strive to maximize internal profit. In such cases, the negotiating process is likely to be long and expensive; the legal costs high; the required performance bonds and guarantees expensive; demands on the transferring firm in terms of personnel and time excessive; the risk of non-payment relatively high; and the effectiveness of international arbitration or national judicial processes doubtful.

The propensity of firms to transfer technology internationally to unrelated parties is related to many factors. One study concluded that:

> ... for companies that *only* evaluate licensing as an alternative to foreign direct investment, the amount of licensing done is greater for companies that (1) spend more of sales on R&D, (2) are relatively smaller in their industry, (3) are more diversified, and (4) have less experience in foreign operations, as measured by the proportion of total sales manufactured abroad by controlled subsidiaries.
>
> Conversely, companies that have no licensing, or negligible licensing, (1) spend less on R&D as a percentage of sales, (2) are among the large firms in their industry, (3) are less diversified, and (4) have a relatively greater percentage of their sales manufactured by foreign controlled subsidiaries.[2]

These quoted observations are limited, of course, to the propensity of firms to transfer technology internationally *via license* externally, that is, to unrelated firms abroad. It is not clear that the findings apply to *all* categories of technology transfer agreements.

Other empirical research has related the cost of technology transfer (and hence the propensity to transfer same) to a number of factors relating to the supplying firm itself, such as:

1. The length of its experience with the technology involved (that is, the technology's maturity vis-a-vis the firm), which means that for the more mature technology, routines have been established, skills developed and defined. Hence, the transfer cost is perceived as less than would be the case otherwise.
2. The size of the firm, in that a larger firm has internal access to specialized staff support for technical feasibility studies, legal counsel, and financial analysis. The cost of any one transfer is thereby reduced. An economy of scope comes into play.
3. The number of prior international technology transfers made by the firm (that is, the overall international experience of the firm), which results in some learning and, hence, lower transfer cost and resistance to transfer. It also leads to the capacity of the firm to self-insure against failure—for whatever reason—in respect to single technology transfers (that is, a "portfolio effect" operates in respect to individual projects, thereby minimizing the down-side risk for any *one* of them).
4. The number of markets in which the firm has existing marketing and distribution facilities. If a firm has a worldwide marketing system for its products, the transfer of technology to an independent firm may jeopardize the transferring firm's position in one or more of those markets. It may create its own competition unless it can effectively limit the area in which the technology may be used or the relevant products sold. The effectiveness of such restrictive covenants obviously rests on both parent and host-country laws on competition. Hence, the transfer cost may be seen as high.

One researcher reports that the cost of international technology transfer is a function of the extent to which the technology is completely understood by the transferring firm. He equates this experiential factor with the number of applications of the technology by the transferor, the age of the technology (that is, the time since its development), and the number of firms utilizing the same or similar technology. One could summarize these three factors as experience, maturity, and level of competition. Presumably all three are correlated negatively with cost of the transfer, and positively with a firm's propensity to transfer. The greater the experience, maturity, and level of competition, the lower the cost of, and the greater the propensity to, transfer.[3]

One assumes that perceived cost of an international technology transfer bears some relationship to its actual cost, and that both influence a firm's tendency to consider selling or leasing its technology to strangers overseas. In fact, there seems to have been very little data collected on the actual cost of technology transfer, possibly because of the accounting difficulties previously discussed. A 1976 study suggested that the technology transfer cost could run between 2 percent and 59 percent of the total cost of a project, with the mean being 19 percent. However, this research was limited to "disembodied technology." Further, it considered only the cost of transmitting and absorbing the information that had to be acquired if capital equipment were to be used effectively, but the technology embodied in the actual equipment was not included as an element of cost. It also would appear that the author was concerned only with internal transfers, and the definition of cost was not clear.[4]

Another study, which defined transfer cost as the direct variable cost associated with the transfer, produced cost figures which ran from 5 percent to 17 percent (or, from 21 to 41 percent if both returns and costs were spread over time and discounted to the present by 15 percent). The variation related to whether the transfer was to an industrialized market economy, to a less developed country, or to a centrally controlled economy (e.g., to a country in the Council for Mutual Economic Assistance [COMECON]). The lowest relative cost was associated with the first category, the highest with the last.[5]

There seems to be general agreement that higher transfer cost is identified with more labor-intensive technology; lower transfer cost with more capital-intensive technology.[6] In most cases, higher training costs are associated with the labor-intensive transfer. In the more capital-intensive transfer, skills are built into the machines. This apparent fact may relate to the perceived reluctance of many firms based in the more industrialized countries to transfer more labor-intensive technology even where the economics of the situation would appear conducive to its transfer. What the firms may be doing is simply avoiding pushing transfer costs up to unacceptable levels.

Although expected cost of the transfer and its perceived risk has much to do with a firm's propensity to transfer technology internationally, some firms consider themselves to be more in the business of transfer, less in the business of manufacturing in their own facilities, than do others. They really do not consider internal transfer at all. Why should this be?

It is very likely that the relative size of a firm's research and development effort, which can run from close to 100 percent of sales (a pure R&D firm, such as Battelle Memorial Institute) to virtually zero (as for a small retail organization), may be relevant. (See Figure 4-1.) For many manufacturing firms the relative size of the R&D investment (say, between 2 and 10 percent of sales) possibly says something about the anticipated technology-in-use. The greater the speed of technological change, the greater the firm's leverage vis-a-vis a foreign client. Buy and pay up, or be denied future generations of technology. On the other hand, if the level of R&D is *very* high, and the rate of technological change *very* rapid (the high-tech case), then the firm can demand 100 percent ownership—and control—and fre-

Figure 4-1. Possible Relationship Between the Propensity of a Firm to Transfer Technology Externally and the Level of R&D (as a Percentage of Sales).

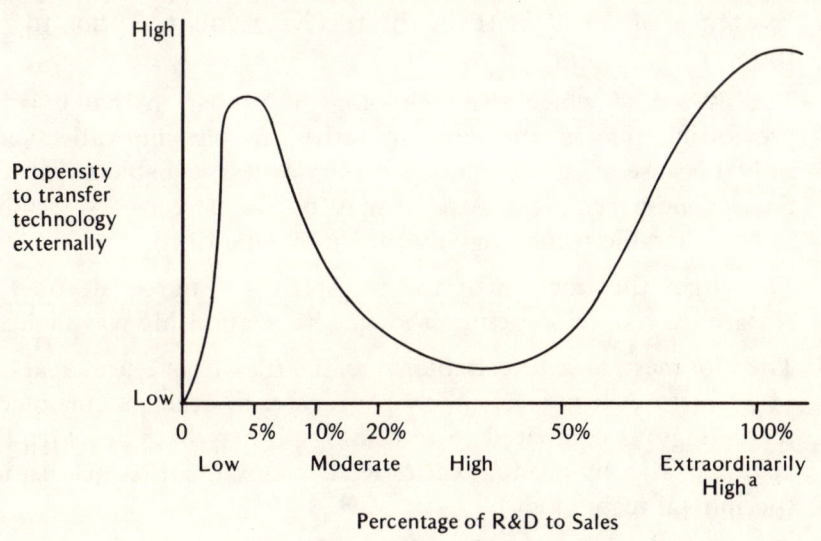

a. Firms which essentially do nothing other than research and development (including the development of people).
Source: Richard D. Robinson.

quently get away with it, that is, so long as the competition either lags far behind and/or is no more flexible in its approach. In such a case, the uniquely advanced technology permits the firm to extract a monopoly rent. Possibly it can maintain that rent most effectively via ownership and dividends rather than license and royalties. Of course, if a host government limits royalties as a percent of sales (say, to 5 percent) and effectively controls transfer prices, but does not so restrict dividends, then the realization of a monopoly "rent" may be possible only through ownership, in which case an internal transfer of the technology is indicated. But even so, there may be exceptions, for dividends and ownership are not necessarily closely linked. (See Chapter 11.)

Davidson and McFetridge[7] have reported that they have found that certain factors seem to relate *positively* to the propensity of a corporation to transfer technology *internally*. These are:

1. The fraction of *internal* transfers made by all companies to the receiving country during the preceding five years (which is possibly a proxy for the host-government attitude toward direct investment; that is, the transfer of unbundled technology as part of the FDI package is freely permitted).
2. Existence of an affiliate in the receiving country prior to the transfer.
3. The degree to which the technology is similar to that existing previously, that is, the more imitative and less innovative (possibly because such technology is seen as less valuable by potential second-party users. Also, it may not be perceived so readily as a marketable technology by the innovating firm).
4. The larger the fraction of the transferor's resources devoted to research and development, although the relationship was slight.
5. The closeness of the technology to the transferor's principal line of business. A firm is less likely to transfer to others its main-core technology as compared to technology it may have developed or acquired as a by-product of its R&D effort or of an acquisition (peripheral technology).
6. The less the prior transfer of all technology by the transferor— which means less experience and possibly exaggerated perceptions of risk and uncertainty by decision-makers.

7. The lower the level of international competition in the relevant technology—which is the argument made above in respect to maximizing monopoly rent by maintaining control of the technology and limiting international transfers to internal transactions.

On the other hand, research[8] has indicated a positive association of the following factors with the propensity of a firm to transfer its technology to *second* parties, that is, to make a transfer via the *external* market:

1. A relatively high investment in research and development as a percent of sales, which relationship seems to be at odds with the finding reported by Davidson and McFetridge, point #4 above, (but note the prior discussion of this subject).
2. The relative size of the firm in its industry, that is, the larger the relative size, the greater the propensity to make external transfers, possibly because it generates a lot of peripheral technology.
3. A relatively high level of diversification by a firm, possibly because it is more likely in such case to be in possession of technology other than that relating to main-line products or processes.
4. The degree of a firm's experience in foreign operations, as measured by the percentage of total sales manufactured abroad by unambiguously controlled subsidiaries—that is, the more limited the firm's experience, the *greater* its propensity to transfer via external channels.
5. The greater the amount of *process* innovation initiated by the firm as opposed to *product* innovation, possibly because of the greater ease of protecting the former from reverse engineering.
6. A relatively high degree of technological competition faced by the firm, which implies more mature technology and, hence, greater labor intensity.

More need be said about #4 above, the degree of experience and its negative relation with the firm's propensity to transfer technology to second parties. (See Figure 4-2.)

It is possible that what is happening here is that the relatively *immature* international firm, one that is growing rapidly in international markets but lacks both the motivation and capacity to exercise centralized control over an integrated international system, is intent on

Figure 4-2. Possible Relationship Between the Propensity of a Firm to Transfer Technology Externally and International Experience (as Measured by the Percentage of Total Sales Manufactured Abroad in Controlled Subsidiaries).

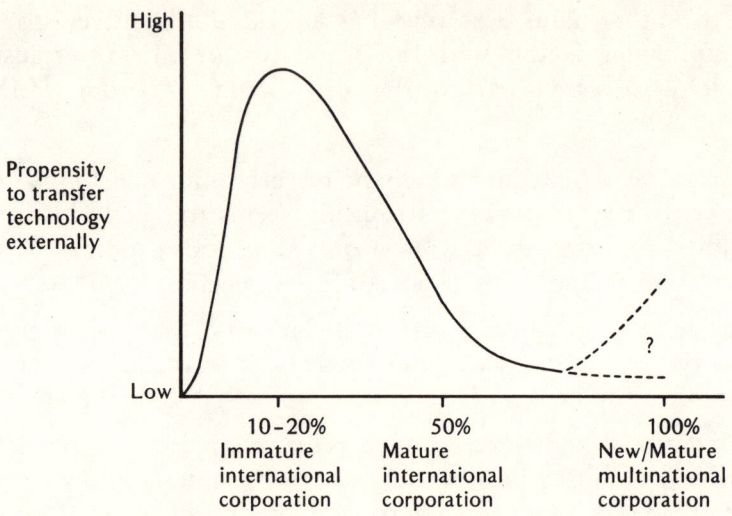

Source: Richard D. Robinson.

entering foreign markets as rapidly as possible and with a minimum commitment of corporate resources. Hence, it has a high propensity to enter via joint ventures and various forms of contractual arrangements—that is, to transfer technology externally. Characteristic of such a firm is a relatively autonomous international division driven by international empire builders. But, as experience is accumulated, and the international division develops the capacity to exercise central control over its foreign operations, management becomes aware of the value of building an *integrated* international system. It begins pulling away from joint ventures and contractual relations to the extent possible and insisting on greater and greater managerial control. Its propensity to transfer technology via license or technical assistance contracts to second parties drops. The firm is now on the way to becoming a true *multinational enterprise*, that is, one which exercises equity-based control over an integrated production system, such control lying in a headquarters corporation that is preeminently owned and managed by citizens of the country in which it is domiciled. The very growth of such a corporation probably lies in its technological superiority over competitors. It is now in a position to extract maximum monopoly "rent" for its technology via internal transfers to its own subsidiaries over which it strives to exercise unambiguous control via majority or 100 percent ownership. That control is believed possible by reason of the experience in foreign markets developed within headquarters, plus the communications and support systems to maintain that control. External technology transfers are seen as less attractive—that is, less profitable.

Researchers have frequently reported that many firms well-established internationally consider external technology transfer only as a last-resort option.[9] Foreign direct investment in wholly owned subsidiaries is nearly always preferred where legally possible and, hence, likewise internal transfer of technology. Given such a universal preference by firms, there is always the suspicion that management is profit-satisfying and power-maximizing. Consider this exchange with an executive in a firm which always insists on doing business abroad via wholly owned subsidiaries:

> Do you ever consider licensing or technical assistance contracts in developing off-shore sources as an alternative to direct investment in wholly owned subsidiaries?
>
> No.
>
> Why not?
>
> We would not have adequate control were we to do otherwise.
>
> Have you ever made a study of the return one might expect from a license or technical assistance contract as a percent of corporate resources committed?
>
> No.

One could well argue that this company did not *know* whether it was profit-maximizing or not. It was also making the assumption that ownership makes possible more effective control over essential elements than does a contract, which I argue in Chapter 12 is a very naive assumption.

In the contemporary world, increased government intervention may, in fact, render it impossible for a firm to enforce the degree of managerial control it would like to exercise on the basis of ownership. Indeed, it can be demonstrated that experienced *multinational* enterprises may now be moving away from control via ownership to control based more on contract, as they learn the ownership spanning an international border is more akin to a contract than domestic ideas of ownership. Limitations imposed on foreign ownership of local assets by both the centrally planned economies (i.e., socialist) and many of the lesser developed countries have forced the multinationals to consider the contract option. And, particularly in a world characterized by uncertain exchange rates, rapid changes in political regimes and laws, and shifting inflation rates, many firms have seen the wisdom of not producing abroad in facilities they own. Being locked into fixed assets, from which it is costly to walk away, may not lead to maximum flexibility and minimum risk. In fact, apparently some firms have found the transfer of technology via contract (including partnership) to be very profitable indeed, given the fact that the corporate investment committed is much less and so, likewise, therefore, is risk.

VALUE-ADDED CHAIN ANALYSIS

The realization of the benefits of transfer by contract has led many firms to analyze their "value-added chain" (see Figure 4-3), so as to ascertain the true nature of their business. It should be borne in mind that as a plant is merely a bundle of processes, a product is a bundle of services, a "value-added chain," if you will. The major links in that chain consist of:

- Information gathering and sorting
- Research and development
- Preliminary production, which is essentially engineering
- Market testing
- Production, that is, hiring, training, managing, purchasing, quality control, financial control
- Market development
- Financing part or all of the process via high-risk entrepreneurial, moderate-risk development, or low-risk working capital for an ongoing enterprise.

In addition, of course, the actual production (what I have called here the application of labor and technical and managerial skills to a series

Figure 4-3. The Value-Added Chain.

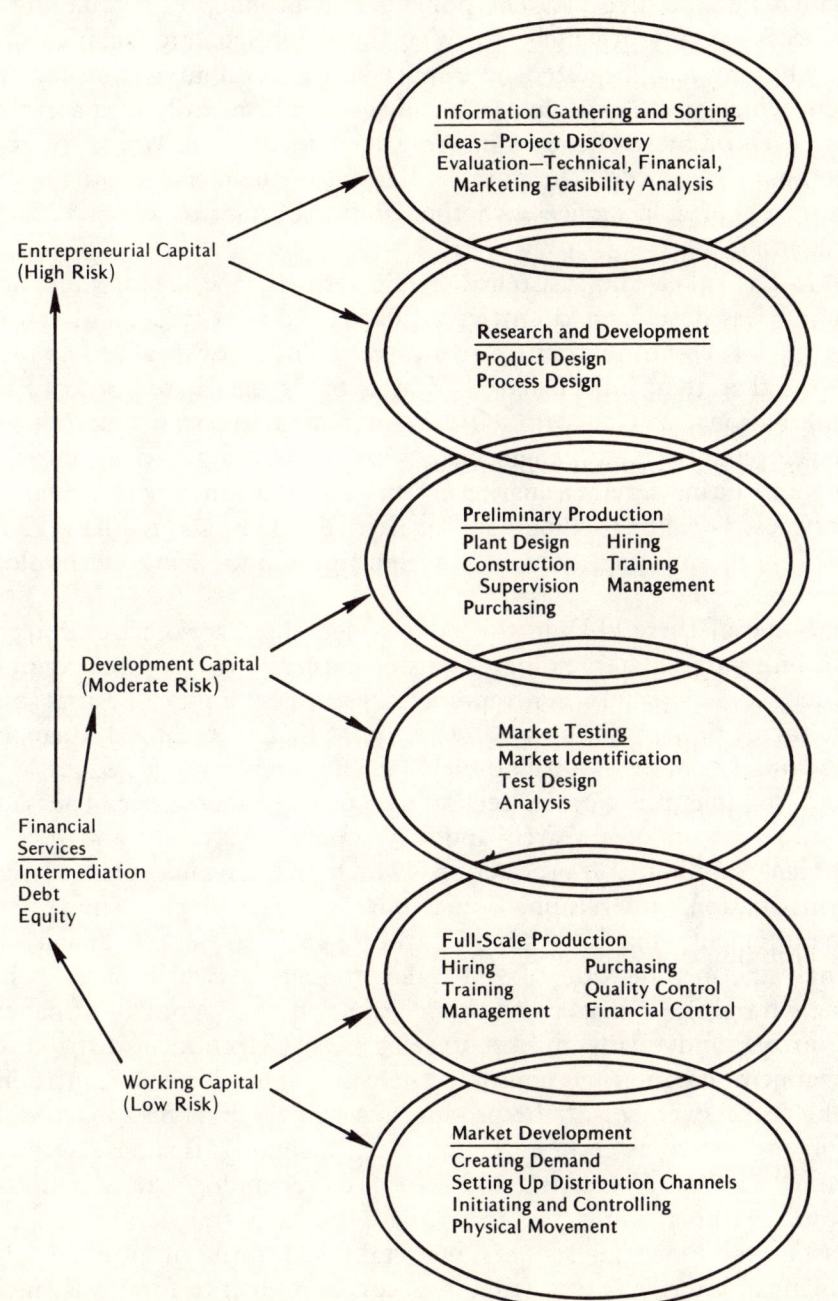

of production processes) is just that—a series of processes, many of which may be divisible. The point is that production is a bundle of processes, many of which can be performed in separate facilities.

The analysis suggested here must be based on an accounting system which gives some clue as to how good a firm really is in performing each of these elements in the value-added chain. Wherein lies its competitive power? The point is that all of them can constitute separate exportable services, whether under contract or as equity or as a loan.

It has taken the insistence by Eastern Europe, China, and now many less developed countries on barter, counter-trade, and various types of co-production to convince many industrialized-country firms that they indeed have products other than hard goods to sell internationally. One can sell entrepreneurial services, research and development services, engineering services, training services, purchasing and hiring services, marketing and distribution services, financial services, as well as selling parts or partially manufactured hard goods for local finishing, which relationship implies some technology transfer.

Some of these links in the value-added chain are obvious in terms of international technology transfer, others less so. For example, how does one transfer entrepreneurial services? Figure 4-3 lists some of them: project discovery and analysis (that is, technical, financial, and marketing feasibility studies), all of which carry an educational and training function, a technology transfer ingredient. The same could be said of research and development (product and process design), preliminary production (which may include plant design, construction supervision, some purchasing, hiring, training, and management), market testing (market identification, test design, and analysis), introduction of full-scale production (which includes hiring, training, management, purchasing, quality control, financial control), and finally market development (selection of distribution channels, creation of consumer demand, initiating and controlling physical movement of the product). All of these services may be the subject of transfers under contract, including financial services, which may also be considered a form of technology transfer in that some learning about the financing process is transferred in many instances. Financial services may take the form of providing (1) financial intermediation (that is, access in respect to foreign financial markets with or without the guarantee of the intermediate, (2) risk-bearing or equity finance redeemable after commercial success (either in part or entirely), or (3) straight debt-financing—or some combination of these. And in all cases, payment may be made either in cash or in goods and services. The point is that these services do transfer technology—the know-how and techniques for entering various financial markets, and hence satisfy the definition of technology transfer suggested earlier.

There is very little hard evidence of the relative profitability of technical transfer via direct investment (that is, internal transfers) and external transfers. A study done by Root and Contractor of 102 licensing agreements produced the results recorded in Table 4-1.

It would appear from Table 4-1 that the gross margin is lower on technology transfer to the LDCs than among the more industrialized market economies, still lower in the COMECON countries. One would hypothesize a greater flow of technology via contract among the more industrialized market economies, more internalized flows to the LDCs. In that many COMECON countries inhibit internal flows, one would expect a much reduced flow generally in their direction. All of these hypotheses seem to be borne out in reality.

Table 4-1. Mean Returns and Costs of 102 Licensing Agreements (*discounted by 15 percent per year*).

	Industrial Countries	Developing Countries	Communist Countries
Total returns to licensor	$411,000	$422,000	$382,000
Direct transfer costs for licensor[a]	85,000	138,000	156,000
Cost as a percentage of return (b/a)	20.7%	32.7%	40.8%
Licensor's gross margin (a-b)	326,000	284,000	226,000
Mean return/cost ratio[b]	35.0%	8.0%	13.7%

a. Only those costs associated with the transfer, exclusive of any contribution to overhead (e.g., R&D) and opportunity costs (i.e., foregone profit on expert sales or return on a direct investment in the recipient's country).

b. The mean of the ratios of the individual agreements.

Source: Franklin R. Root and Farouk J. Contractor, "Negotiating Compensation in International Licensing Agreements," *Sloan Management Review*, Winter 1981, p. 28.

There is evidence that few managers act so as to maximize the return on technology transfer because they fail to consider either the value of the technology to the recipient or the profits from exports or from direct foreign investment that might have been realized had the technology transfer not taken place—that is, to opportunity cost.[10]

NOTES

1. David J. Teece, *The Multinational Corporation and the Resource Cost of International Technology Transfer* (Cambridge: Ballinger Publishing Company, 1976), p. 54. His data suggest that "the average increment to total project cost is about 5 percent for transfers to joint ventures, about 9 percent for transfers to government enterprises in which there is no equity participation by the Transferor" (pp. 80–83).
2. Piero Telesio, "Foreign Licensing Policy in Multinational Enterprises," Unpublished DBA dissertation, Harvard Graduate School of Business Administration, 1977, p. VI-37.
3. Teece, *supra*, pp. 51–53.
4. *Ibid.*
5. Farok J. Contractor, "The Profitability of Technology Licensing by U.S. Multinationals: A Framework for Analysis and an Empirical Study," *Journal of International Business Studies*, vol. 11, no. 2, Fall 1980, p. 46.
6. See David J. Teece, "Technology Transfer by the Multinational Firm," *Economic Journal*, June 1977, p. 242.
7. William H. Davidson and Donald G. McFetridge, "International Technology Transactions and the Theory of the Firm," *Journal of Industrial Economics*, March 1984, pp. 253–64; reported earlier by Davidson, Amos Tuck School of Business Administration, Dartmouth College, Working Paper No. 106, 1981.
8. David Zenoff, "Licensing as a Means of Penetrating Foreign Markets," IDEA, vol. 14, no. 2, Summer 1970, p. 297; and Telesio, *op. cit.*, pp. II-11, 11, 17, 18, and 20.
9. An example is Piero Telesio, *Technology Licensing and Multinational Enterprises* (New York: Praeger, 1979).
10. Franklin R. Root and Farok J. Contractor, "Negotiating Compensation in International Licensing Agreements," *Sloan Management Review*, Winter 1981, p. 30.

CHAPTER 5

A Summary of Relationships on the Supply Side of International Technology Transfer

In order to pull together all of the relationships suggested in the preceding chapter, it may be useful to summarize them and then to build a model portraying those relationships, which—let it be noted—relate almost exclusively to the supply side of international technology transfer. (See Figure 5-1.)

To begin with, a private firm's propensity to transfer technology by whatever route—bundled with the direct foreign investment package, embodied in exported hardware, or transferred separately under contract (unbundled)—is undoubtedly related to management's awareness of a foreign demand, the firm's history, and its resources. More generally, one may state that its propensity to transfer technology is a function of the perceived cost, the perceived risk, and the anticipated benefit to be derived from the transfer, these perceptions being very much derived from its awareness, history, and resources.

Going back one step, one should realize that *perceived cost* is very likely related to (1) the firm's prior relationship with the recipient organization, (2) the public/private ownership mix of the recipient, (3) the firm's experience with the technology, (4) the absolute size of the firm, (5) the number of prior technology transfers made by the firm (by whatever mode), (6) the labor/capital intensity of the technology (perhaps a proxy for complexity), (7) the availability of external financing for the transfer (e.g., governmental), and (8) the adequacy of the firm's accounting system.

54 / THE SUPPLY SIDE

Figure 5-1. Relationship of Factors on the Supply Side of the Technology Transfer Process.

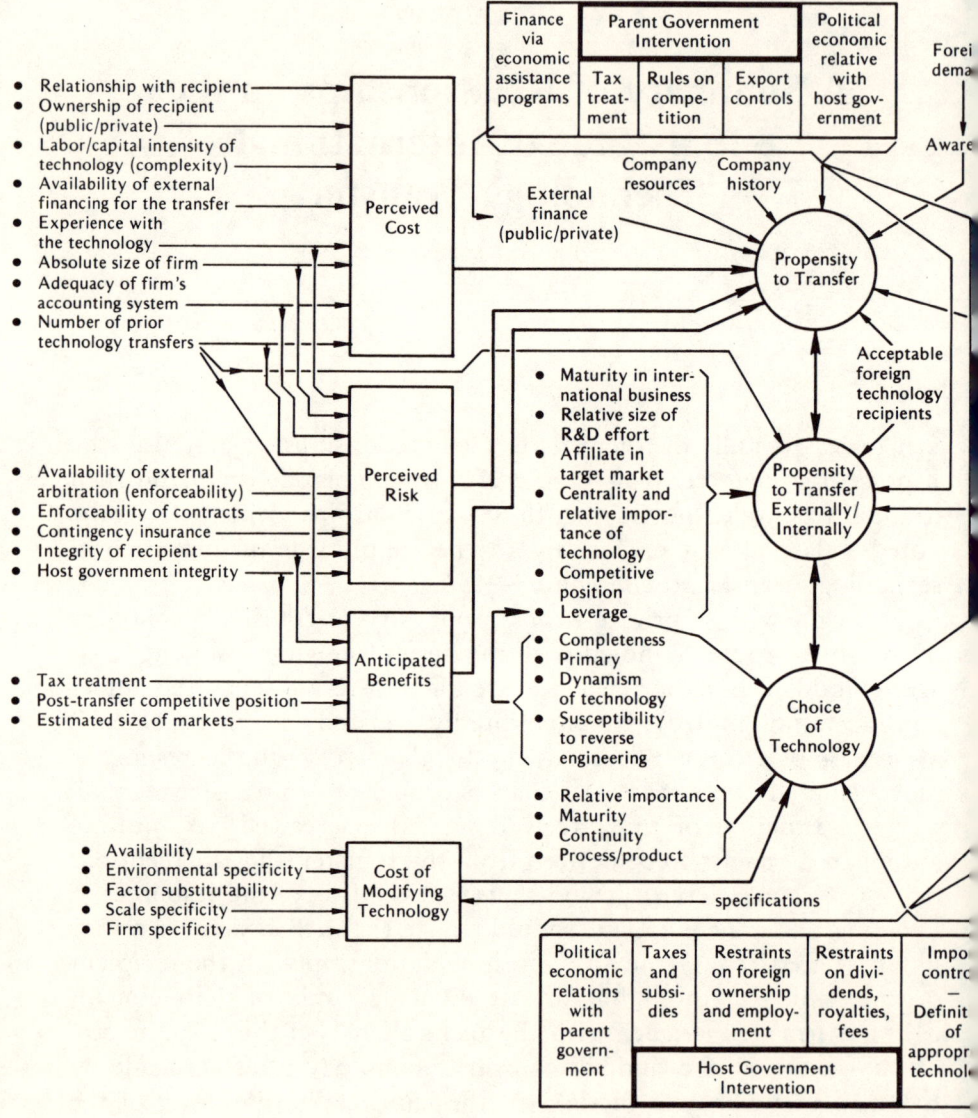

Perceived risk, on the other hand, is probably also a function of (1) the firm's experience with the technology, (2) the absolute size of the firm, and (3) the number of prior transfers, but one should also add five other factors. These are: (4) management's level of confidence in the integrity of the receiving organization and (5) of the host government, (6) the enforceability of contracts, (7) the availability of external arbitration (and the enforceability of arbitration awards rendered abroad), and (8) the availability of contingency insurance (such as commercial and/or government-backed guarantees against non-payment of earnings, expropriation of assets—including contractual rights—and losses due to war and insurrection).

Anticipated benefits also rest on (1) the adequacy of the firm's accounting system, (2) perceived integrity of both the receiving organization and (3) its parent government, as well as (4) the tax treatment of different flows of earnings (dividends, earnings on export sales, fees, and royalties), (5) the competitive position in which the transfer will leave the transferor after the transfer (which again relates to the adequacy of the firm's accounting system), and (6) to the estimated size of markets.

In the U.S. case, there is a special tax factor which becomes relevant to these considerations. Since 1981, all expenses for R&D conducted in the United States are allocated to U.S.-source income, even though the principal benefactors may be overseas affiliates. This rule provides a way of reducing taxable income in the United States, particularly if the technology flowing from this R&D is transferred to foreign affiliates at little or no cost to the recipient. The U.S. parent firm might well transfer technology to its own subsidiaries abroad without any license or technical assistance agreement calling for payment of royalty or fee. The full benefit will be captured in any event by the U.S. parent via the flow of dividends. However, dividends need not be repatriated immediately, whereas royalties and fees are usually paid on a regular basis. Hence, there is more opportunity in the dividend case to use foreign earnings for reinvestment purposes, deferring the payment of U.S. taxes until those earnings are actually remitted to the United States. The firm gains to the extent of the interest on the U.S. tax it would otherwise have had to pay during the period of the deferral. If the foreign tax is low, or has been waived as a tax incentive, this gain may be considerable.

Therefore, the firm may be encouraged to do R&D within the United States rather than transfer that function abroad, thereby elevating foreign profits and reducing domestic profits. Of course, an inbound transfer of technology under contract could result in the parent's payment of a fee or royalty to a foreign affiliate, which payment could constitute a deductible expense against U.S. taxable income, but this gain is likely to fall far short of the full cost of the R&D expenditure, which, in the first case (i.e., domestic R&D) may be deducted from current taxable income.

Up to the enactment of the Tax Reform Act of 1986, there had been an even more favorable tax situation which had the impact of encouraging firms to conduct R&D within the United States. It was possible, given certain circumstances, to treat the transfer of technology, even to a subsidiary—whether the technology is patented or not—as the transfer of a capital asset, thereby permitting the royalties and fees received to be treated as capital gains, not income. In such event, of course, the flow was until 1986 taxed at a reduced rate by the U.S. government. That difference is now eliminated, which may have the effect of encouraging U.S. firms to conduct R&D abroad. There is now no tax reason for not doing so.

In any event, induced by a favorable combination of tolerable perceived cost and risk and an attractive benefit, the firm is faced with a decision as to whether to effect an internal or external transfer, assuming that both opportunities are legally and commercially viable. Foreign governments sometimes intervene to block internal transfers via direct foreign investment, or transfers embodied in imported goods. The market may be limited by law or decree to firms transferring unbundled, disembodied technology via contract. On the other hand, likely candidates to receive technology, or those willing to pay what is considered by the supplying firm to be a "fair" price, may not always be present in the market. Indeed, the host government may enter into any negotiation and so limit the price and other conditions as to make the transfer unattractive to the transferor (or *potential* transferor). The role of governments is dealt with in some detail in a subsequent chapter.

But let us assume that the firm is realistically faced with the choice of an internal or external transfer. What seems to influence choice at the first level is the firm's relative maturity in international business, the size of the firm's R&D effort (very large R&D investment tends to be associated with external transfer; in the modest range, with internal transfer; in the low range, with external transfer)—which is the reason possibly that research has not seemed to provide an unambiguous, one-directional relationship in the study reported earlier. Other factors influencing the choice between internal and external transfer probably include the firm's prior technology transfer experience, its ability to protect the technology from unauthorized use, the existence of an affiliate in the market, certain characteristics of the technology involved (specifically, its relative importance and centrality), the firm's competitive position vis-à-vis its technology, and its leverage. If a firm is faced with many firms offering the same or similar technology on the market in unbundled, disembodied form, it may feel compelled to do likewise.

Leverage, or the ability to maintain a tolerable level of control over the technology and the recipient even in the absence of special legal devices is probably a function of four characteristics of the transferred technology—its completeness, its primacy, its dynamism, and its susceptibility to reverse engineering. How complete is the transfer? How primary is the transferred technology? How dynamic is the technology, that is, how rapidly can it be expected to become obsolete? How easily can it be replicated? The point is that a firm is likely to become less and less inclined to transfer technology to second parties as the transferred technology becomes more complete, more primary, is less subject to rapid obsolescence, is more central to its principal business, and is more subject to reverse engineering.

Given a firm's propensity to transfer technology, either internally or externally, the selection of a particular technology rests largely on a match between the perceived specifications and the cost of modifying existing technology to meet those specifications. This cost, in turn, is a function of the availability of something resembling the specified technology, degree of environmental specificity, factor substitutability, scale specificity of existing technology, and the extent to which the technology is specific to the firm.

We are now in a position to chart the relationships of these factors on the *supply side* of the technology transfer process for a private firm.

PART III
THE DEMAND SIDE

CHAPTER 6

The Recipients

We now switch vantage points and look at the technology transfer process from the point of view of the recipient.

As on the supply side, the entities involved may be either a private enterprise seeking to maximize internal financial results, a government enterprise motivated at least in part by the external or social impact of the transferred technology, or a non-profit organization carrying out certain explicit goals. Within the latter category fall educational institutions, local charitable organizations (whether of a religious or secular nature), and independent research establishments. These non-profit organizations may be funded from purely local sources or have access to some international support, as are the Rice Institute in the Philippines and the Wheat Institute in Mexico. Certain financial intermediaries that facilitate the international transfer of technology should be mentioned in this context, such as the many regional development banks, the World Bank and its affiliated organizations, and the special agencies of the United Nations.

Finally, there are a variety of essentially bilateral conduits assisting specific countries in the identification of technology, the modification of that technology, the acquisition of technology, and the facilitation of its absorption. One could mention the U.S. AID Program, the U.S. Peace Corps, the Swiss-based Institute for Intermediate Technology, the United Kingdom-based "Columbo Plan," the

European Communities' Economic Development Fund, and the like. There is no point in going into detail here, for their number, scope, purpose, and manner of operation are constantly undergoing change. As before, our principal preoccupation is with the private-enterprise sector, but with some comment on state or parastatal enterprise.

It can be assumed that a firm's appetite for foreign technology rests on the costs, benefits, and risks that it perceives to be associated with the transfer. To the extent these differ from those faced by the transferring (supplying) firm, they are of interest, though there is an obvious symmetry to a degree. Even so, perceptions may be at considerable variance.

A foreign firm's wholly owned local subsidiary, even though its management operates with a high degree of autonomy, may be reluctant to accept the transfer of new technology from its foreign parent if such transfer is likely to erode the subsidiary's autonomy— or it is *perceived* that it is likely to do so. The cost is not a financial one in this case, but rather social-psychological, which comes with an enhanced feeling of dependency and captivity.

Likewise, a government-owned enterprise may admit to a higher internal financial cost in reference to a given technology transfer, but still maintain that, by its elaborate and time-consuming (and, hence, costly) process for selecting a technology and monitoring its transfer, it is, in fact, safeguarding external economic/social interests to an extent that a purely private enterprise cannot and does not. The externalities of a new technology are likely to be part of both the cost and benefit calculation of a state enterprise, less so by a private firm unless forced to consider such by law or regulation.

The externalities should be seen as variables intervening between government rules and regulations and their implementation. Such externalities can result in either economies or diseconomies, and may be classified under several headings: local value added, impact on balance-of-payments, public revenue effect, growth-generating influence, income/wealth distribution effect, innovational effect, and impact on political development. A paragraph on each will suffice for our purposes here. Bear in mind that all of these measures may impact on the definition of what constitutes "appropriate technology." (The devices a government may introduce so that transferred technology is more likely to produce the desired impact is treated in a later chapter.)

EXTERNAL IMPACTS OF TECHNOLOGY TRANSFER

Value-added has to do with the *net* effect of a technology on the commercially valuable productivity of a country's land, natural resources, labor, capital assets, and financial capital. Obviously, value-added in this sense does not simply equal the sales revenue of a firm, less the value of what it purchases from others. The labor, capital, and materials a firm uses may already be employed in producing something else. To derive a net value-added, that something else has to be deducted. Also, any activity of the firm induced by a new technology which impacts positively or negatively on economic activities outside the firm should be included in this calculation, such as the training of persons who will leave the firm's employ to use that training elsewhere in the country, or the stimulation of greater efficiency in supplying industries due to increased demand, or inducing new or improved production "down-stream" toward the final consumer by reason of providing a new, better, or cheaper intermediate product.

Anything which generates greater or more productive employment of existing resources—including people—adds to value. However, the *maximum* employment of one particular resource may not lead to *maximum* value-added. For example, if the employment of the last person does not add enough value (marginal product) to cover the cost of subsistence for that individual, then the country would presumably be better off—in an *economic* sense—in substituting another technology (some form of mechanization), even if the unemployed individual could be kept alive only by public relief. Governments have been known to intervene to prohibit the import of labor-saving technology in the face of massive unemployment, such as a refusal by the Turkish government to permit the import of cotton-picking machinery some years ago.

Balance-of-payments effect refers to the net influence the use of a new technology has on a nation's annual cash flow balance with the rest of the world. A positive flow could occur if the technology leads—on balance—to a more efficient use of scarce resources and thus releases exportable resources (or resources usable in producing exportable resources), the lowering of service or product costs (thus making a local industry internationally competitive), or to the development of a new *internationally competitive* industry which either earns or saves foreign exchange.

The *public revenue effect* of a project involving a technology transfer has to do with the balance between the taxes and fees an enterprise pays to public authority and the cost of publicly financed services used, such as personnel trained in public institutions, subsidized utilities, security services, public transportation, port facilities, etc. used by the enterprise. Much depends upon the tax structure of a country. If it rests heavily upon property taxes, then that technology requiring heavy capital investment is likely to have the highest positive impact on public revenues, at least in the short run.

The phrase *growth-generating effect* in respect to a technology has to do with the stimulation of a country's economic growth over time. In many poorer countries, the bulk of the population has a very short time-horizon. The use of resources in such manner as to provide immediate improvement in consumption may be seen as highly desirable. Future income is discounted heavily. Saving, rather than consuming added income, and investing in such ways as to improve *future* income (and consumption) may not be very appealing to the local public. But the political elite in such a country, which is very likely to have satisfied much of its *own* basic needs, is inclined to have a longer time-horizon and be committed to long-term economic growth of its country at a fairly rapid pace. A policy of austerity—the rationing of consumer goods—may well result, although the problem of providing enough improvement in consumption so as to maintain incentives must be borne in mind. For example, one technology may have to do with the manufacture of cosmetics, another with the production of pharmaceuticals. The former may be very much more profitable; the demand is high, and, unfettered, the industry would expand rapidly. The result is that the net value-added is high—for a time. The pharmaceutical technology, by making certain medicinals more readily available (i.e., at lower cost) to the local population may add to the health and vitality of the people and may be more supportive of *long-term* economic growth, even though the enterprise generates a less attractive, internal financial result. We say "may be" because one cannot know without looking at a specific situation very closely. The point is that the short-term profitability of an enterprise incorporating a particular technology, even though there may be substantial value-added in the short run, does not necessarily identify that technology with the greater long-term growth effect.

Distributional effect relates to the impact of a technology on the distribution of wealth and income in a society. For example, very capital-intensive technology is likely to concentrate the distribution of income to the owners of the capital assets, and possibly to the managers if they are sufficiently insulated from owner control, as in a large corporation with thousands of stockholders, none of whom hold unambiguous control of the firm. A labor-intensive technology is more likely to result in a somewhat broader distribution of benefits, although not necessarily if a "sweatshop" type of exploitation is socially and politically tolerated.

Innovational effect is a somewhat more subtle concept. Economic growth, which seems to be a general objective of all societies, albeit under certain restraints, ultimately rests on continuing innovation or change—cultural, social-psychologically, and technologically. The relevant question here is whether the transfer of a particular technology is supportive of constructive innovation in the recipient country to the maximum extent possible, given the resources available. Of course, innovational push can be so powerful as to be destructive to orderly development. Costly social-political conflict can result as traditional ideas and values are challenged too rapidly. But with that proviso aside, the relevant consideration here is whether or not a technology transfer is such as to induce continuing innovation in the host society. A technology which requires the training of many local nationals in its application, particularly if local market conditions require modification of the transferred technology, will possibly induce this innovational push, particularly if the transfer induces major changes in *user* technology and the users are many and/or constitute an influential element in the host society, an element whose behavior will be emulated by others. The innovational impact of a technology tends to become greater as the *transferred* technology becomes more "primary" and each transfer more "complete" (in the sense that we have used the terms in Figure 2-2).

The *political impact* of a technology transfer has become evident during the preceding discussion. Power relationships within a society are obviously heavily influenced by changes in relative economic power generated by a technology transfer. Wealth and income distribution may be altered between economic groups, geographical regions, countryside and metropolitan areas, and the private and state (political) sectors of a society. The processes of industrialization and urbanization to which much technology transfer tends to lead may radically change political power balances. The rapid development of certain groups in a society through the application of technology may well challenge traditional leadership. And if a minority ethnic-religious group has been the principal beneficiary of technological transfer, internal conflict could be the result.

APPROPRIATE TECHNOLOGY

As suggested earlier, all of these concepts can be wrapped up in the phrase "appropriate technology," a phrase which is only meaningful if defined within the context of a society's economic and socialpolitical priorities. Over time, such priorities are either explicitly or implicitly determined by some sort of political process, however centralized or decentralized that process may be. No objective function can be applied to the allocation of resources—technology among them—until these priorities are established. It is all very well to say that a technology is appropriate if it produces a good at a competitive price, but where markets are very imperfect, that is not much of a guide. Society may be providing a hidden subsidy by providing resources at lower-than-market prices or extracting a hidden tax by doing the reverse.

Presumably in a relatively poor country, where labor is relatively abundant and capital very scarce, the most appropriate technology is that which uses the most labor and the least capital, *provided* that such technology turns out a product acceptable to the market. If that market be an external one (international), a quality standardized within quite narrow limits may be required, a tolerance which labor-intensive technology may find it difficult or impossible to meet.

There are other provisos as well. One is that maximum economic growth may be society's object, not maximum employment. The point is that the genesis of economic growth is increased productivity of an entire economy, which may best be achieved by a somewhat more capital-intensive technology demanding a higher order of local motivation, perception, and skill. Committing virtually everyone to low-skilled labor in traditional industry is not exactly conducive to change—that is, change in attitudes, way of life, values, motivations, perceptions, skills, and finally, in economic well-being. At this point, the notion of "intermediate technology" becomes germane. This phrase refers to that stratum of technology lying between the traditional, labor-intensive technology and the most complex, state-of-the-art technology.

Another proviso is that traditional, labor-intensive technology may be appropriate only if the society does not feel threatened by external foes and feels no need to develop a modern military sector and the capital-intensive skills and industry to support it.

Other provisos required, if the usual definition of appropriate technology is to hold, are: (1) that the society does not feel impelled to improve the quality of its human capital through education and measures to prolong life; (2) that the society has no concern for conserving its natural resources or protecting local ecosystems; (3) that the society does not feel obliged to direct resources to develop or support underprivileged parts of its population or its more depressed regions; (4) that labor-intensive technology is available for all tasks society wants done (which depends on defining the task, e.g., mail delivery or instantaneous communication). One could probably think of more provisos underlying the equation of appropriate technology with labor intensity.

In the final analysis, appropriate technology is that technology *which does what society wants* at least cost, given the total labor hours of input, to create both the capital invested in the production technology and in the actual production process. If the technology is imported from abroad, then we must include the capital invested in the transfer process and in developing adequate local absorptive capacity.

THE COST OF RECEIVING TECHNOLOGY

The absorptive capacity of a society has much to do with the cost of international technology transfer. In one situation the cost may be relatively low—a situation in which local people have already mastered the relevant skills (the result of large and long-standing social investment in education and training), where enterprises capable of applying the technology exist, where the necessary inputs (raw materials, energy, water, space, component parts, etc.) are readily available at reasonable cost and tolerable quality, and where the political environment is such that costs can be calculated with some degree of accuracy over the near- and intermediate-term future. On the other hand, in many situations one or more of these conditions may not exist, which means that the receiving enterprise must internalize some or all of these costs via in-plant training, vertical integration, relocation, and/or the setting up of special contingency funds to hedge against politically inspired changes in taxes, regulations, fees, etc. Under those circumstances, the cost of technology transfer to the recipient firm can be relatively high.

There is some evidence that a country's absorptive capacity vis-a-vis imported technology relates positively to per capita gross national product, manufacturing as a percent of gross domestic productivity, and the absolute number of scientists and engineers in the country.

The point has been made that the higher a country rates against these measures, the more it is inclined to import proprietary rights without associated technical assistance when receiving technology. Lower ratings mean that it is inclined to buy fewer proprietary rights and more technical assistance. For example, one researcher reported that for U.S. licensors, the percentage paid for proprietary rights of all payments for technology transfers stood at 91.2 for West Germany, 5 for Nigeria. That is, for every dollar paid U.S. licensors by Nigeria, only $.05 was in payment for rights while $.95 went for assistance.[1] The data was from 1975 and only for technology under "license" to affiliates, which makes any generalization suspect. Still, it is highly plausible that the less developed a country is, the more it is concerned with improving its absorptive capacity via technical assistance. Proprietary rights are likely to be less relevant. Indeed, much of the transferred technology may be of an older generation, not the state-of-the-art, and hence not subject to legal protection via patent or secrecy warranting a license.

This last point brings up the subject of transferring second-hand capital equipment and associated technology. Many countries, in their drive toward modernization, insist that only the most up-to-date technology be transferred, which rules out the import of second-hand equipment in many cases. This option fell into some disrepute in years past when foreign firms attempted to capitalize in their foreign affiliates used equipment as though it were new. Current practice is for the recipient government to insist that any used capital equipment be assessed by independent industrial machinery assessors and that such evaluation be certified by its local embassy or consulate.

There is fragmentary recent evidence that the international market for used capital equipment is growing as its advantages become known and accepted. In many cases, the embodied technology is likely to be more appropriate in the sense that it is somewhat more labor-intensive, is more closely akin to technology currently in use, requires fewer skilled technicians not available locally or who cannot be easily trained, is more established (i.e., mature), and less risky (that is, the "bugs" have been worked out), can be used to produce goods acceptable in international markets, and finally, is substantially cheaper.

> When a durable product such as a machine, appliance or automobile wears out, most of us simply discard it and buy a new one. But there is an alternative; have the products remanufactured. This approach, used more often behind the scenes in manufacturing than most of us realize, is destined for a greater role as energy and materials become scarcer.[2]

The Center for Policy Alternatives at the Massachusetts Institute of Technology has undertaken the only comprehensive study reported to date of the benefits of remanufacturing and of the conditions under which it makes sense. The Center reported that virtually any product or machine is a possible candidate for remanufacture. "The major requirement is that there be a discarded product, called a core, in which the cost of salvaging the material and value added is much less than the market value of the remanufactured item." Obviously, certain conditions must exist. The product must be capable of being disassembled. There must be a continuing demand for the product; that is, the embodied technology must be relatively mature and significant change not anticipated. The production of a product satisfactory to customers in terms of performance must be possible. In the United States, it was found that extensive remanufacturing was going on in respect to automotive parts for older model vehicles, industrial equipment (for example, hydraulic equipment, heavy-duty diesel engines, production metal-working machinery), commercial products (examples being office machinery, compressors, communication equipment), and residential products (such as power tools, lawn mowers, and appliances).[3] One suspects that the products of the remanufacturing industry are finding their way increasingly into international trade both as exports and as the substance of capital investment.

An important example of the remanufacturing industry involving the international transfer of technology lies in the paper machinery industry. For a variety of reasons during the 1970s, it became apparent that the remodelling and updating of existing paper-making machinery was more economical than the purchase of new machines. The rebuilding process had to be undertaken within the plants of the paper manufacturers by teams of engineers and technicians from the paper machinery makers' organizations, not infrequently located abroad. Hence, considerable technology transfer took place as the customers' own employees were informed and trained by the rebuilders.

A subject infrequently introduced in reference to the cost of receiving and absorbing technology is the extent to which the ingredients, dimensions, and performance of a transferrable technology have been standardized in the society to which the transfer is contemplated, and how compatible those standards are with those used by the major trading countries. Transfer may be facilitated or rendered more difficult; lesser or greater modification is required. Some countries apparently feel that the acceptance of foreign standards, albeit internationally recognized, is somehow subversive to national sovereignty and is evidence of intolerable "dependency." It may be useful to list a few categories of the thousands of existing standards, by the way of example. It is said that some 10,000 standards are required for the tolerably efficient working of a modern industrial society. Examples:

1. The metrification of physical dimension—height, width, weight, area, distance, volume, etc.
2. The threading of screws and bolts
3. The sizing of lumber, metal sheets and bars, bearings, newsprint, books, doors, windows, tires, electrical connections, light bulbs, wire mesh, gears, cable, memory chips, etc.
4. The calibration of time, pressure, temperature, voltage, amperage, wattage, caloric content, etc.
5. The dimensions of road vehicles and railway rolling stock
6. Radio and television frequencies
7. Sizes and compatibility of records, discs, cassettes, films, etc.
8. Electrical voltage (for household and for industrial use)
9. Horsepower and braking power

10. Measures of toxicity of products and acceptable means of disposal of toxic waste
11. Measures of environmental pollution and emissions
12. Occupational safety
13. Allowable additives and contaminants in food and pharmaceutical products, fuel, water, chemicals, etc.
14. Industrial robots
15. Computer hardware and software
16. Safety standards for boilers, vehicles, aircraft, gas pipelines, fuel storage tanks, electrical transmission, etc.
17. Building standards—homes, industrial buildings, high-rise, bridges, docks and piers, breakwaters, etc.
18. Professional qualifications
19. Liability for product malperformance, environmental damage, etc.
20. The grading of fruits, vegetables, lumber, textiles, etc.

One could go on almost endlessly.

The point is that a firm located within a society applying standards used generally in the more industrialized countries, the source of most technology being transferred internationally, is advantaged compared to a firm situated in a country lacking generally applied standards or using standards not compatible with those in general use internationally. For the latter, the cost of ingesting foreign technology may be quite a bit greater than would otherwise be the case.

In response to the obvious need for establishing international standards over a wide range of goods and services, continuing effort has been mounted to bring such standardization about, both on the regional level (for example, the European Communities, the European Free Trade Area, and the Economic Commission for Europe, which includes both Eastern and Western Europe), and on the global.

In the 1960s, three European countries formed a regional electrical certification system to facilitate European trade in electrical products. The system was closed to non-member countries. U.S. manufacturers, for example, wishing to sell their products (or technology) had to secure certification in each individual country. This type of problem led to the multilateral Agreement on Technical Barriers to Trade, popularly known as the Standards Code, which was first negotiated and accepted during the Tokyo Round of Multilateral Trade Negotiations between 1975 and 1979 under the auspices of the General Agreement on Tariffs and Trade (GATT). This code establishes international rules among governments regulating the procedures by which standards and certification systems are prepared, adopted, and applied, and by which products are tested for conformity with standards. The code also specifies non-discriminatory treatment of products and technology of different national origin, equal access to certification systems, use of international standards wherever possible, a preference for formulation of standards in terms of performance rather than design criteria, and a prohibition against using standards and certification systems to create unnecessary obstacles to trade.

Additionally, there is the International Organization for Standardization (ISO), a specialized agency of the UN, to achieve international standardization, membership of which consists of the national standards bodies of some 100 countries. It works through literally hundreds of technical committees, the work of which is coordinated by a secretariat based in Geneva, Switzerland. During its life, which began in 1946, the ISO has brought about the harmonization of some 5,000 or more standards. It works in close cooperation with a variety of other organizations, important among which is the International Electrotechnical Commission (IEC), which was the first international organization in the standards field (established 1906). The IEC deals with all electrical products and focuses on five main categories:

1. Common means of expression: vocabulary, graphical symbols for electric circuit diagrams, units, and their associated letter symbols, etc.
2. Standard methods of testing and specifying performance, thereby facilitating comparison of claims made regarding quality of performance and the specification of minimum requirements
3. Definition of acceptable levels of quality and performance as a result of the standard testing
4. Features affecting mechanical or electrical interchangeability or aimed at reducing the variety of models
5. Safety standards

Many of these international agreements also require notification to other countries of any new standards that are introduced.

From the point of view of those involved in the international transfer of technology, the problem related to standards is a fourfold one:

1. The existence of divergencies in national standards
2. Wide variation of national measures for the enforcement of standards
3. Quality assurance systems based on limited country participation
4. Ignorance on the part of technology supplier and/or recipient of the relevant technical regulations

The existence or non-existence of technical standards is but one factor relevant to the costs perceived by the firm proposing to receive foreign technology. One must also consider such factors as a firm's prior experience with international technology transfer, prior experience with the relevant technology, and past relationship with the technology supplier. All obviously influence the perceived cost of the transfer in that experience should produce a learning effect by reducing the transfer cost. Likewise contributing to the perceived transfer cost are such factors as firm size, firm ownership, the complexity of the technology to be transferred, the ability of the firm to lay off part of the cost of the transfer (and risk) to another entity, such as a government financial intermediary, and various restrictions imposed on the technology transfer by the transferring entity. Each of these factors is discussed below.

Firm size is related to the pool of resources available to the firm—financial and human—and to the organizational "slack" existing within the firm. Slack in this sense refers to resources which, although presently employed, could be shifted to other assignments without materially reducing efficiency. On average, the larger the firm, the lower the perceived cost of a technology transfer. Its cost of capital may be lower. It is more likely to have resources that could be freed up and applied to the technology transfer and absorption process (i.e., "slack"), not the least of which is managerial. The transferred technology may be more closely akin to that used presently someplace within the firm. Hence, a relatively large firm size possibly has the net effect of reducing technology transfer cost to it, although one notes exceptions.

Ownership of the firm may rest in a family, a limited number of associated owners, the general public, or the government. In general, it is probably safe to assume that the more diversified is the firm's ownership, the less costly a given technology transfer is perceived. Why? In a family firm, management and ownership are often indistinguishable. Introduction of a new technology may be perceived as threatening control by the family-managers because of their lack of personal knowledge of, and mastery over, the new technology. The cost is, therefore, seen as very high, for it would mean essentially turning control of the firm, at least in part, over to "outsiders." As one moves away from that model, the loss of personal managerial control by owners becomes less likely. Professional managers are employed. Of course, if the firm were controlled to a significant degree by employees, as in the German "co-determined" or Yugoslav "self-managed" enterprises, the introduction of new technology may be stoutly resisted by reason of the perceived cost in the loss of jobs, or at least in their redefinition in unacceptable terms. Government-owned enterprises are possibly the most likely to react to the external costs ("diseconomies") occasioned by the transfer of a new technology. Some of these have been discussed in preceding sections.

A recipient firm is very likely to face higher absorptive costs for more complex technology (more capital- and skill-intensive technology) if the technology represents a discontinuous change for the receiving firm, thereby rendering its present skills and capital equipment largely obsolete. However, it should be borne in mind that the transfer of highly complex technology may also be relatively easy, *if* the new skills employed are largely built into the capital equipment and only a handful of individuals carrying new skills are required. On the other hand, the laws and regulations of many countries, LDCs as well as many of the more industrialized countries, make the rapid introduction of more capital-intensive technology very costly in that the firm may be held responsible for the continuing employment of the employees rendered redundant by that technology. At the very least, heavy penalties may be imposed.

Another important element in the recipient firm's perception of cost is the presence of any restrictive covenants in the technology transfer agreement. The transferors of technology typically try to impose upon the recipients various restrictions or conditions related to the use of the transferred technology. Some of these restrictions are:

- Tied-buying provisions, which may be hidden in unnecessarily tight specifications and linked to some degree of quality control by the transferring firm. Such provisions may have the effect of virtually requiring the technology recipient to buy certain materials, components, machines, or continuing services from the transferring firm, not in the open market, even though the underlying agreement does not *explicitly* require such purchase. In such cases, price becomes a key issue, as well as quality and delivery.

- Technology grant-back, often at no cost to the transferring firm, of any innovation, new adaptation, or improvement achieved by the recipient of the transferred technology. But, one might well ask, precisely what constitutes "new" technology as compared to innovative, adaptive, or improvement technology? How does one distinguish such in a completely unambiguous way?

- Export restrictions which specify that the sale of a product manufactured at least in part with the transferred technology may not be sold in specified countries, possibly none outside the recipient's home market. Such restrictions obviously place the relationship in a collision course with most host governments, which are intent on maximizing exports. The restrictive provisions may require permission of the technology transferor before export, permit exports only to certain countries, prohibit exports altogether, limit exports to the technology supplier's own agents or distributors, or require that the supplier's name or trademarks not be used on exported products (or possibly the reverse; they must be used).

- Application or field-of-use restrictions, which stipulate that the transferred technology may be used only in certain applications or fields (a group of related uses). An example of this restriction is the use of a process in the manufacture of motor vehicles but not for the production of aircraft, even though it might be equally applicable.

- Exclusivity provisions prohibiting the transfer of the technology to third parties without the express, written permission of the transferring firm. Such a provision can run counter to a host country's explicit objective of disseminating useful, new technology throughout the country as rapidly as possible. A restriction of this nature may even outlive the validity of the underlying transfer agreement.

- Prohibition on collateral licensing or technology transfer such that the technology recipient may be obligated not to obtain similar or complementary technology from other technology vendors in the same field. The primary reason for such a restriction is to reduce the likelihood that valuable information would be divulged to third parties. The fear is that the second technology supplier might have access to the recipient's facilities.

- Commitment not to contest the validity or ownership of any proprietary information or rights transferred. Let it be noted that under the laws of many countries, the United States included, such a commitment may be held illegal and unenforceable because any rights and obligations embedded in a contract are co-determinous with the life of that contract.

As one study pointed out:

> It is important to note that in many cases restrictions adversely affecting the trade and development of developing countries [and adding cost to the receiving firm] may not be explicitly stated in contractual arrangements. This may be true especially in those cases where formal contractual arrangements are not required or called for in view of the control which may be exercised by the investing firm [or the technology supplier] over the local firm, be it a subsidiary or an affiliate company in the developing country. Restrictions on production and export may well be applied over the activities of the subsidiary or affiliate.[4]

Therefore, it may be argued that *internal* technology transactions are likely to be more costly than external transactions to both the receiving firm and to its parent country. In external transfers, restrictive covenants are, generally, explicitly exposed and can be evaluated. On the other hand, the degree of control a foreign management in fact exercises over the use of transferred technology by a subsidiary or affiliated local company may not be at all clear. Both organizational conflict and local law and regulation may restrain such control. That subject is examined in more detail in a subsequent chapter.

The availability of external financing at special rates for the transfer of technology from abroad may constitute an important cost-reducing factor from the point of view of the recipient firm. Parent governments, in pursuit of national development plans, may offer subsidized financing to encourage local firms to acquire foreign technology in specific sectors. Indeed, for a variety of reasons the parent government itself may acquire the technology and make it available to interested local firms.

Some of the reasons inducing such government intervention is the relative small size of many markets served by individual local firms;

they may be too small to interest potential technology providers. Also, an individual local firm may not have access to adequate information relative to alternative technologies and sources, nor possess the necessary technical skill to evaluate technological alternatives. Moreover, if there are a number of local firms in the same field requiring similar technology, all may import technology for similar purposes from different sources and different countries, "which would result in the importing of a multiplicity of technologies without adding significantly to the technical capacity within the country," and not necessarily at the lowest cost.[5] Fragmentation of the technology market, and reduced scale, may well affect the price adversely from the technology receiver's point of view.

RISKS AND BENEFITS

Perceived risk by a recipient firm obviously varies with the size of the firm (a large firm can absorb a failure), the complexity of the technology (the degree of its novelty to the recipient), and the availability of external financing and the conditions under which it is given. If the financing is of such a nature that repayment may be waived in case the technology cannot be used in a commercially successful way, the risk inherent in the transfer is clearly less than might otherwise be the case. Likewise, if the risks in an international technology transfer (non-delivery, non-performance, incomplete transfer, undue dependency upon the source, or an ineffective transfer—possibly due to inadequate training provisions or ducumentation), the availability of legal recourse to collect damages or the provision of specific guarantees may be of critical importance in the recipient's perception of cost.

Several approaches to this subject need to be mentioned. First, the technology recipient may schedule payment in such a way that the final 10 to 20 percent, or more, is withheld until a satisfactory transfer is assured. An alternative is to require that the purveyor of the technology post some sort of performance bond which becomes payable in whole or in part upon the failure of the technology supplier to perform its obligations under the technology transfer contract. A local bank may arrange through its correspondent bank in the country of the technology supplier to hold a performance bond that is callable if and when the supplier fails to perform. To protect the interests of all parties concerned, an arbitration process should probably be set up in advance so that if differences of opinion arise as to what constitutes material non-performance under contract, such differences can be resolved with relative ease. It is standard practice for an arbitration clause in an international technology transfer contract to specify arbitration in a third country, to refer to a body of interpretive law acceptable to both parties, and to commit both parties to some established international arbitration process, such as that offered by the International Chamber of Commerce or by the United Nations Commission on International Trade Law.

Anticipated benefits of technology transfer to a technology recipient are essentially a function of the anticipated impacts of the new technology in terms of reducing unit cost, improving product quality, or introducing an entirely new product. But whether the anticipated benefits are of real significance depends upon the size of the market, the price-demand elasticity of the relevant product, and the availability of technology otherwise not available and with which the firm would soon be competing in its markets. There is always the question as to whether the technology might better be developed via local research and development rather than purchasing or leasing it from abroad. That choice is based on calculations as to relative cost. Of course, if the local environment is such that the imported technology will require substantial modification before application, then some local R&D may be required in any event. Of particular importance is the degree to which (1) the technology is environmentally specific along some of the dimensions already discussed, (2) factor substitutability is possible, and (3) it is scale-specific. The perceived cost of modifying a technology will clearly impact on the choice of technology, assuming that there are any alternatives.

PROPENSITY TO SEEK FOREIGN TECHNOLOGY

The propensity to seek foreign technology on the part of a potential receiver of technology from abroad is, as is true on the supply side as well, related to its perceptions of cost, risk, and benefit. Also relevant is the firm's awareness of a desired and available technology abroad, plus its propensity to seek an external transfer (as opposed to an internal transfer via a merger, joint venture, or partnership with a foreign firm). Much depends upon the size of the firm (and of its market), the complexity of the technology (degree of novelty to the firm), and upon the extent of a firm's relevant internal R&D re-

Figure 6-1. Relationship of Factors on the Demand Side of the Technology Transfer Process.

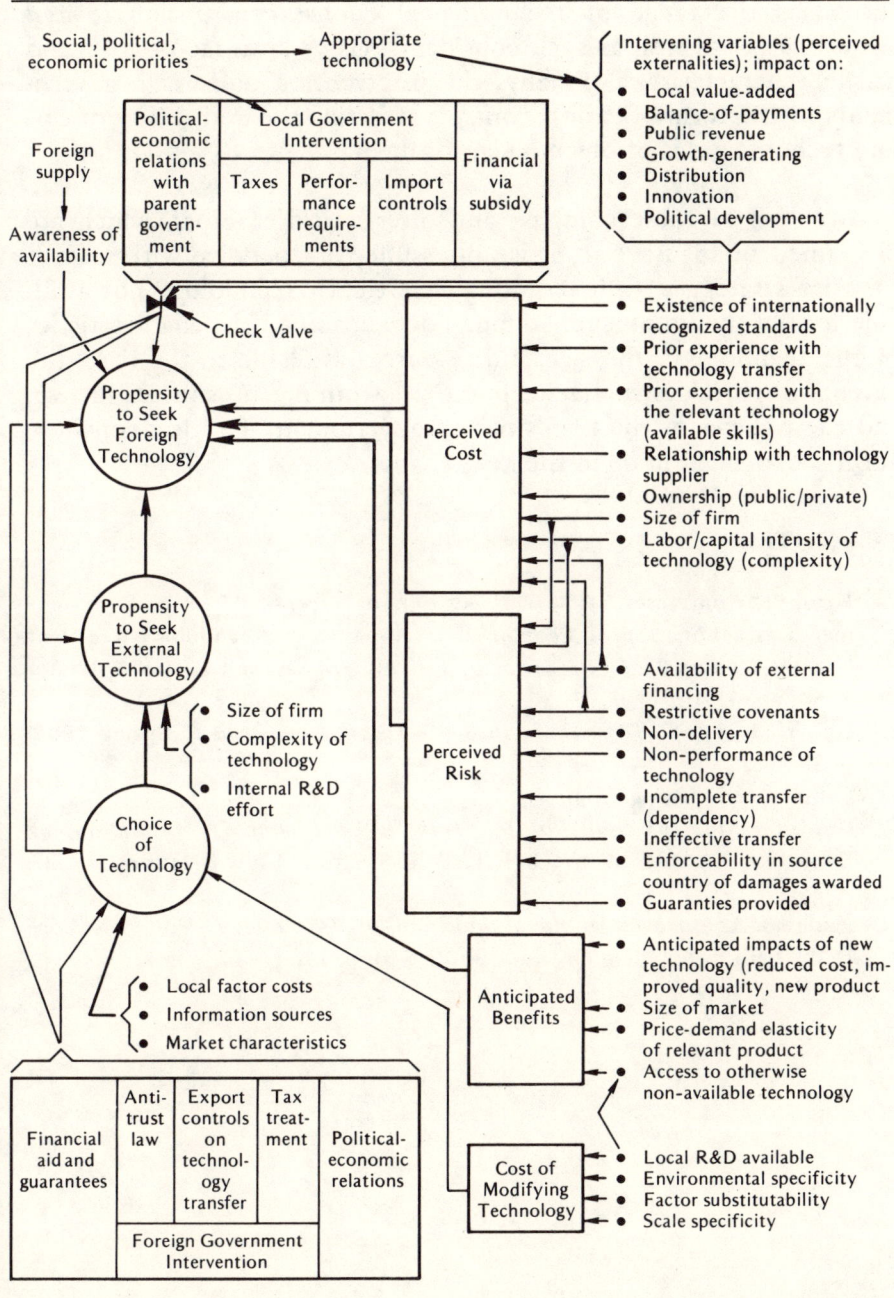

sources—skills, management, facilities, funds. Also impacting on a firm's propensity to seek technology via external channels from abroad is the range of technological choice confronting it—*and known to it*—as well as such considerations as local factor costs and market characteristics. Finally, the intervention of foreign governments in the form of export controls affecting the availability of certain technologies is of obvious significance.

Given all of these complex and inter-relating factors, which are diagramed in Figure 6-1, it is impossible to generalize with respect to what strategy a firm requiring a particular technology not available locally should pursue. So much depends upon the characteristics of the technology, the size and resources available to the firm, the extent to which alternative technologies from diverse sources suffice, and the policies of the firm's parent government. This last consideration is expanded upon in the next chapter.

NOTES

1. Farok J. Contractor, "The Composition of Licensing Fees and Arrangements as a Function of Economic Development of Technology Recipient Nations," *Journal of International Business Studies*, vol. 11, no. 3, Winter 1980, pp. 47–62.
2. Robert T. Lund, "Remanufacturing," *Technology Review*, Spring 1985, p. 5.
3. Summarized from *Ibid.*, pp. 5–13.
4. *Guidelines for the Study of the Transfer of Technology to Developing Countries*, a study by the UNCTAD Secretariat (United Nations, 1972), p. 23.
5. *National Approaches to the Acquisition of Technology* (United Nations Industrial Development Organization, 1978), p. 14.

PART IV

SPECIAL ISSUES

CHAPTER 7

Responding to Intervention by Governments and Multigovernment Organizations

The assumption underlying the following discussion is that the commercial success of an international technology transfer is greatly influenced by the intervention of governments, singly or collectively. In this discussion, we deal only briefly with the *specific* forms of intervention—the laws and regulations and codes introduced by individual governments or multigovernment organizations—in that they are constantly changing. Rather, the focus is more on the general nature and impact of such intervention. Two relevant subjects—the protection of proprietary rights and the international bidding process—are the subjects of separate chapters and not included here, although they might well have been in that both involve forms of government involvement.

SOME EXAMPLES

The parent governments of the technology supplier and the technology recipient both may actively intervene in ways which influence the perceived costs, benefits, and risks associated with international technology transfer. In recent years, the intervention of multigovernment organizations has become of greater importance. We consider first the parent government of the technology supplier, typically a more industrialized country.

Industrialized Country Interventions

Many countries, particularly those active politically, periodically impose trading sanctions against other countries. Such sanctions include technology transfer whether the technology is embodied in hard goods or moves in "disembodied" form via written or orally conveyed information, data, or skills. Sanctions may arise from strategic defense considerations (as for COCOM[1] countries), out of foreign policy considerations (as vis-a-vis South Africa), or by reasons of domestic short supply or the stockpiling of strategic materials. In the U.S. case, a firm may be compelled to enforce U.S. strategic- or foreign policy–based sanctions on foreign enterprises which the U.S. firm controls. The definition of what constitutes control is a bit vague, but presumably even if the U.S. firm has a minority equity interest in a foreign venture, but had contracted to supply "strategically important" materials and/or technology on an *ongoing* basis, it would be compelled to shut off that supply regardless of contractual obligations. It would be deemed to have effective control over the flow. For materials, technology, and know-how already transferred, it would have no control, unless the U.S. firm had a distribution contract with the foreign venture. If, on the other hand, the U.S. firm had sufficient control of the foreign venture to block the sale of product and/or technology from the venture to the countries under sanction, then it would be obliged to do so under U.S. law. Other COCOM member countries have similar regulations.

The taxation of income received from a technology transfer clearly affects calculation of net earnings. Note that the ability of a technology supplier to use a foreign tax credit against its parent-country tax liability may be limited by an ownership test and otherwise influenced by treaty relations. For example, for a U.S. firm to credit foreign income taxes paid by an associated foreign firm against its U.S. tax liability, the U.S. firm must own at least 10 percent of the equity of the foreign firm. If not, foreign taxes may only be expensed (and thereby reduce taxable income), but not credited. In negotiating an international technology contract, it may be desirable for the transferring firm to negotiate payment as a percentage of the recipient firm's earnings and, hence, put itself in the position of being able to claim a domestic tax credit.

Antitrust law or rules of competition enforced by the parent country of the technology supplier may also be relevant to the transfer. In general, under U.S. law for example, attempts by the U.S. supplier to restrict the use of a transferred technology may be held illegal unless those restrictions are based on a valid proprietary right—patent, trade secret, trademark, or copyright—which right automatically bestows on the owner the legal right to restrict use under many circumstances. However, if these rights are purely ancillary to the basic transfer agreement containing the restrictions, they may be ruled illegal, even when such a ruling has the effect of restraining competition, or potential competition in trade within the United States or in foreign trade of the United States, either export or import. Likewise, a joint venture or partnership (collaboration) between two firms, even when both are based outside the United States, may be considered illegal if the undertaking reduces actual or potential competition in U.S. foreign trade or within the United States. A joint venture, partnership, or a simple contract between a U.S. firm and a foreign firm could likewise be illegal if the latter is, or could reasonably be expected to become, a competitor in the United States and is prevented by the relationship from becoming such. If, however, the foreign enterprise is "unambiguously controlled" by the U.S. firm (e.g., if the U.S. firm owns over half of the foreign firm's voting stock), then the relationship is probably all right, even though none of this makes much sense, given the fact that ownership and control need not be related and that control is a many-faceted function. The laws relating to competition of no other country or region (e.g., the European Community) have the extraterritorial application as does U.S. law in this regard.

Less Developed Country Interventions

Turning to constraints imposed by the laws and regulations of many less developed countries relevant to international technology transfer, one is struck by the enormous variation. All, of course, are designed to protect the interests of the host society, it being assumed that deals struck via enterprise-to-enterprise negotiation may not be in the national interest. One reason for this assumption, as discussed elsewhere, is that prices faced by the venture may not measure the true scarcity value of the goods and services to be used or produced by the venture (e.g., foreign exchange, capital, labor, infrastructure services, etc.). Prices may be set by fiat, subsidized, or inflated by a tax. A second reason is the presumption that negotiations between unequals—a large, multiproduct, geographically diversified firm versus a smaller, single-product, geographically concentrated firm—is unlikely to lead to an equitable arrangement for the latter, which is likely to be the party on the LDC end.

Limitation of foreign ownership to under 50 percent is imposed by many governments (as in Yugoslavia); in others, only if the venture is to enjoy some benefit (e.g., tariff-free access to a regional group, as in the Andean Common Market [ANCOM]), or if the enterprise is located in a specific sector (e.g., public utilities or "strategic" industries in many countries). The imposition of such a limit, particularly in high-risk ventures, means that local investors must carry substantial risk. Also, when limitations are first imposed, preexisting ventures in which foreign investors have over 50 percent ownership are caught unless there is a "grandfather clause" exempting them. More frequently, these older ventures are required to comply over a stipulated period of time. It should be noted once again that a minority foreign-equity interest does *not* mean that the foreign partner or technology supplier lacks effective control over one or more aspects of the venture's operations. It all depends upon who has leverage over whom in terms of market access, continuing technology supply, skills, and capital availability. But many firms facing such a limit on foreign ownership simply opt out of a market.

The legally required reduction of foreign equity in a venture over time to a specified maximum percentage (e.g., 10 percent over ten years, as in Indonesia in many cases) enables local investors to avoid the high-risk period of a venture's start-up and initial growth, but it discourages foreign investment in the first place, and, hence, possibly technooogy transfer as well. Given the risk-bearing entrepreneurial function carried by the foreign investor, it must anticipate a higher flow of returns than would be the case if it could retain its ownership share for a longer period. Such a heightened threshold rate of return rules out many projects which might otherwise be perfectly feasible. The transfer of technology may be seen as too risky for the anticipated return. If the fade-out takes place over a relatively long period, say twenty to twenty-five years, then the question arises as to what strategy the foreign investor will start pursuing (or try to) in year fifteen or twenty. It may stop transferring all new technology, desist from any new investment, repatriate all possible funds (including depreciation allowances)—in short, start liquidating the venture and thereby leave only a shell. Or, if some formula guaranteeing fair compensation for the relinquished equity were promised (e.g., based on expected earnings), then the foreign investor might inflate earnings via inadequate depreciation and inappropriate transfer pricing. In either case, the local joint-venture partner could end up being exploited, that is, paying more than the foreign share is really worth.

In order to induce some risk-sharing by the foreign partner, a few governments require a *minimum* equity participation by the foreign partner (e.g., 25 percent in the case of China) if the foreign partner is to participate in earnings at all. An alternative policy would be to withhold valuable incentives unless a minimum foreign equity involvement were exceeded. In either case, such a requirement would seem to be considerably more rational than an *upper* limit, in that a minimum participation induces some sense of responsibility on the part of the foreign investor unless, of course, its other transactions with the joint venture are seen as the major source of profit, not dividends.

As suggested above, certain incentives may be withheld by the host government from a joint venture unless *local* participation exceeds a given minimum (e.g., 25 percent in Saudi Arabia to enjoy tax incentives). Such a policy carries a certain rationale in that if local resources are to be used to subsidize a venture, which is what a tax reduction implies, local entrepreneurs and investors should be given the opportunity to participate. A tax exemption is, of course, tantamount to free use of publicly supplied infrastructure—roads, sewer and water systems, police and fire protection, etc. Such conditions obviously influence the mode of international technology flow seen as desirable.

Many LDC governments impose performance requirements on ventures involving foreign direct investment or the ingestion of foreign technology, thereby inhibiting the ability to negotiate an agreement freely. Such performance requirements may include commitments to export a given percentage of output (or value of same), produce specified products at a given volume via specific technology on a specified site, employ a given number of local citizens at various levels, capitalize the firm above a specified minimum, increase the local value-added according to a fixed schedule (phased manufacturing), etc. Each such restriction increases the risk perceived by the parties involved, particularly if the penalties for non-observance are not clearly specified. Hence, the cost of doing business is increased. Anticipated payouts must be higher than would otherwise be the case so as to cover the heightened risk and uncertainty.

Multigovernment Interventions

Various international and regional bodies (for example, the Commission of the European Communities and the specialized agencies of the United Nations) have become increasingly active in the effort to influence the initiation and use of technological innovation. These activities include the control of international shipments of hazardous waste, the imposition of an environmental impact assessment requirement by public, international financing institutions; the control of trade in products involving endangered species; imposition of industrial safety codes; guidelines on the shipment, production, and use of hazardous chemicals; curbs on the international sale of certain ecologically dangerous agrochemicals; and the creation of consumer-rights guidelines. The problems inherent in such approaches are (1) the application of *nationally* developed standards, via trade controls and the requirements imposed on trade and investment by national financial institutions, to other countries whose interests may be quite different, and (2) the growing maze of disparate national laws and regulations. Adding to the problem is the fact that codes issued by the specialized agencies of the United Nations have no real force unless enacted into national law or inserted into contracts.

It would appear that some technology transfer is motivated in part to avoid national regulation. A firm may be motivated not only to minimize criminal charges or monetary penalties imposed by governments in certain situations, but also liability which might arise to consumers and the general public. Although possibly the most spectacular product and hazardous waste liability cases have been American, the concept of product and waste liability seems to be spreading quite rapidly to other industrial countries. Although spreading somewhat less in third-world countries, they are not uninvolved.

One of the most widely publicized cases, leading eventually to UN involvement (1981), related to the marketing of an allegedly "inappropriate" product on LDC markets. The product was infant formula, of which the Nestlé Corporation was a principal supplier. The result was a worldwide consumer boycott campaign during the late 1970s against Nestlé products and the approval by the UN World Health Organization (WHO) of a code for the marketing of infant health food. The situation had to do with the alleged misuse of infant formula by consumers and the marketing practices pursued in the third world by Nestlé and other producers of like products. These practices allegedly had the effect of misleading consumers—or, at least, not providing adequate warning of the dangers of misuse. A variety of issues surfaced, such as the degree to which governments were responsible for the policing of local marketing practices, the extent to which manufacturers should be held responsible for the use of their products contrary to explicit instruction—in this case, according to a specific mix ratio of formula to water, the use of pure water, use only in sterile containers, and the general superiority of breast-feeding over bottle-feeding. In other words, the case related to what I referred to in Chapter 2 as "user technology."

The anti-Nestlé spokespersons claimed that the firm had acted to enhance sales unconscionably by providing free samples to hospitals, by rewarding salespeople for volume of sales achieved, and by under-emphasizing the dangers of use in largely illiterate poor countries where the firm should have anticipated that the instructions would or could not be read, that the water used would be polluted, that the containers would be far from sterile, and that the formula would be diluted well beyond what would constitute a useful and sustaining food for infants. Hence, the firm was culpable. On the other hand, the Nestlé management insisted that in many places it was required by local governments to supply free samples to hospitals and clinics, where mothers were encouraged to use bottle-feeding rather than breast-feeding by local health personnel. Although the anti-Nestlé boycott ended with the alleged acquiescence of Nestlé to the WHO marketing code, the issue has not been resolved.

Another famous—or infamous—case involved hazardous industrial emissions, specifically the Union Carbide-Bhopal tragedy in which several thousand were killed and tens of thousands injured by reason of an apparently accidental release of a deadly chemical into the air (or possibly, as some alleged, a deliberate act of sabotage). Union Carbide had been cited previously by the Indian government as a "good citizen" by reason of the fact that it had trained Indians to operate the plant, had turned over its management to Indians, had permitted Indian investors to take up nearly 50 percent of the equity in its Indian subsidiary, and generally had tried to abide by all Indian laws and regulations. No personnel in the employ of Union Carbide U.S. were on the premises. The safety equipment and procedures in the plant had been analyzed two years before the accident by Union Carbide personnel. A report had been issued shortly thereafter pointing up certain deficiencies. The plant was subject to periodic safety inspections by Indian government personnel. It developed after the accident that not all of the safety measures required by Union Carbide had been implemented by the local Indian employees. Litigation and negotiation continue.

Several important, and as yet unresolved issues, surfaced in the subsequent and ongoing discussion. First, could Union Carbide, the parent U.S. company, be held responsible? Had not the Indian government assumed at least some degree of responsibility by inducing Union Carbide to turn over the plant entirely to local personnel and by periodically inspecting the plant? The actual owner of the plant, Union Carbide's Indian subsidiary, enjoyed separate corporate identity from Union Carbide U.S. and was owned differently in that over 49 percent was Indian-owned. Did liability go through to the parent corporation? In any event, how did one measure liability—by U.S. or by Indian standards? Did one accept the notion of punitive damages and contingency-based legal fees, both acceptable under U.S. law, but not under Indian law? Obviously, such events and their ultimate disposition could have major impact on where potentially dangerous materials are produced and, hence, influence directly the international transfer of technology. In fact, an effort is under way within the United Nations to create an international safety code for the chemical industry. The Commission of the European Communities is likewise attempting to define company responsibility in this area.

A book on international technology transfer should not ignore efforts by the United Nations to draft a "code of conduct for technology transfer," a code which would take into account the legitimate rights of proprietary technology owners and the needs of developing countries' technology buyers. In May 1974, the UN General Assembly adopted the "Declaration on the Establishment of a New Economic Order." An accompanying program of action assigned a high priority to a code of technology transfer. A draft code in 1977 recognized the right of governments to regulate international technology transfers within the framework of relevant international law, treaties, and agreements. In so doing, the draft listed a number of practices considered to be unduly restrictive, such as requiring payments after expiration of industrial property rights, exclusive grant-back provision, the waiving of challenges to validity of a property right, restricting the freedom of the technology buyer in respect to the use of similar or complementary technology from other sources, restrictions on research, restrictions on use of personnel, price-fixing, restrictions on adaptations, tying arrangements (requiring purchase of equipment, materials, and know-how from the technology supplier), export restrictions, restrictions on publicity, and requirements imposed on the technology-acquiring party to provide equity capital or to accept managerial participation by the supplier. Even though not yet cast in a final and universally approved form, the draft's provisions alert governments to some of the problems inherent in international technology transfer.

GENERALIZATIONS

In general terms, a government has only four administrative levers with which to encourage or discourage an enterprise, whether foreign or domestic, to do something it would not otherwise do, including the international transfer of technology. The four levers: regulation. tax incentive, subsidy, and reduction or increase of uncertainty.

A government is inclined to intervene positively (that is, provide incentives) when it perceives, rightly or wrongly, that the *external benefits* expected from an enterprise's activities will not be reflected in its internal financial results within the time relevant to a firm's decisions. Or, the government's objective may be to prevent an enterprise from doing something that it would otherwise do, but which the government deems socially undesirable in the sense that the associated *external costs* are not reflected in the firm's internal financial results within the time frame of decisions. The assumption in either case is that an enterprise can be expected to respond only to the expected financial consequences of an activity, not to external costs and benefits. In such cases, the government may impose either incentives or disincentives. It is useful to consider the characteristics of these four administrative levers.

Definitions

Regulation is a government device to induce an enterprise—whether it be local, foreign, or a joint undertaking—to do something it would not otherwise undertake on its own volition (or vice versa). Regulation will work only if the firm perceives that, even though the act required (e.g., external technology transfer) may be less profitable than the prohibited alternatives (e.g., direct investment), it is nonetheless sufficiently attractive in terms of anticipated financial results to warrant undertaking. If the enterprise has much to lose in the form of "sunk" cost (expenditures already made and which cannot be recovered, as in the development of a specific technology), or is threatened with the loss of a significant market possessing the promise of future profit, then this approach may be effective—*but* at the possible cost of discouraging subsequent investors or technology suppliers (new entrants). The discouragement will be particularly intense if the regulation is viewed as arbitrary and changeable (unpredictable) and without compelling rationale. Uncertainty is thereby increased. Compliance, of course, is mandatory for the enterprise choosing to enter the market.

Tax incentives imply a tax reduction (or increase) from that normally imposed on inputs, outputs, and/or financial results (profits), provided the enterprise satisfies certain conditions (or fails to do so). Compliance with these conditions is voluntary. That is, the enterprise need not comply if it is willing to accept the tax consequence. (In a sense, regulation is equivalent to an infinitely high tax.)

A *subsidy* is the provision by a government of specific services or other inputs to an enterprise at below-market prices, such as a grant to encourage local R&D activity or provision of below-market interest on export financing, etc. A subsidy should really be viewed as a continuum from what amounts to a concessionary ("subsidized") price, an outright gift ("grant"), to an actual payment. That is, included under the general heading of "subsidy" are inputs provided at specially discounted prices (market price less some percentage), at zero price, or at a negative price. The latter, of course, results in a payment to the enterprise, such as a transfer of funds for conducting research and development locally, for maintaining employment at a given level, or as payment to the enterprise for its output at a price above the market price (government procurement at a premium price). This subsidy continuum may be represented as in Figure 7–1.

As in the case of a tax incentive, the firm has the option of complying with the qualifying conditions or not. If not, it simply foregoes the subsidy.

Figure 7-1. The Subsidy Continuum.

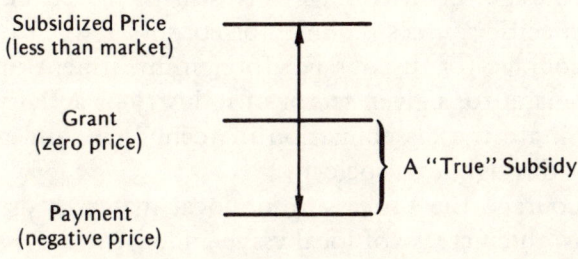

Note: A "true" subsidy is often identified as a payment, either to pay the firm for doing something that it would not otherwise undertake, such as research and development, or to reimburse the firm for certain expenditures. Also included would be the provision of some service at no cost, for example, the identification and evaluation of a foreign technology, possibly even the negotiation of terms under which it becomes available to a local enterprise, or the reverse—free government assistance in finding a foreign recipient for a particular technology.

The *reduction of uncertainty* as perceived by a management has the result that a firm's anticipated financial return need not be discounted to the same degree as might otherwise be the case. Uncertainty may be reduced by a government guarantee, such as a minimum royalty guarantee, but only if perceived by management to be enforceable. Examples of uncertainty reduction are (1) a commitment by a government to purchase part or all of the product of a project at a known price, (2) assurance of a measurable market by prohibiting further entry by other firms, (3) rendering technology transfer agreements enforceable under international law by permitting binding external arbitration, (4) entrance into international agreements in respect to the payment of compensation in the event of expropriation of assets or breach of technology transfer contract, and (5) participation in international conventions for the protection of proprietary rights.

The purposes of all four devices—regulation, tax incentive, subsidy, and reduction of uncertainty—are similar. Generally, they are introduced:

1. to encourage (or discourage) the local investment of foreign savings;

2. to encourage (or discourage) development of a production facility (or use of a particular technology) on a particular site;
3. to encourage (or discourage) investment or technology transfer from specific sources (country or corporation);
4. to encourage (or discourage) foreign investment in, or technology transfer to, a given sector of industry or activity;
5. to facilitate the dissemination of technology and relevant skills widely throughout a society;
6. to encourage the processing of local materials prior to export (that is, the increase of local value-added);
7. to encourage the use of locally produced components so as to induce the development of ancilliary industry or to discourage the use of materials in short supply;
8. to encourage the use of local labor;
9. to encourage (or discourage) the transfer of foreign technology of certain types;
10. to encourage the development of certain functions locally (such as data processing, research and development) or to discourage others (toxic waste–polluting, highly dangerous, etc.);
11. to encourage the transfer of foreign skills (whether technical or managerial);
12. to discourage the emigration of technically skilled persons;
13. to encourage the development of exports or to discourage them in the event of politically imposed sanctions;
14. to discourage unnecessary imports;
15. to encourage the transfer of ownership, control, and technical-managerial skills to local nationals;
16. to encourage the public sale of equity in an enterprise;
17. to encourage lower prices on outputs and to prevent monopolization of markets;
18. to encourage production to certain standards (such as employee health and safety, product quality, emissions and toxic waste, etc.).

In all four cases, a government may limit application of a specific incentive or lever with respect to nationality (domestic or foreign), to undertakings only above a certain minimum size, to the utilization or transfer of certain categories of technology ("appropriate," advanced, etc.), or to a certain point in the development of an undertaking.

In regard to the last—the *timing* at which a regulation is imposed, or tax incentive or subsidy given, or an uncertainty-reducing guarantee provided—it is useful to consider the life cycle of a project, which may be diagramed as in Figure 7-2. Bear in mind that some activities may be conducted simultaneously and others added; much depends on the nature of the project.

Obviously, the further removed an activity is from a predictable result (the point at which aggregate investment is recaptured, plus a reasonable profit, say at Point Z in Figure 7-2), the greater the uncertainty perceived by the parties involved in an enterprise. Early on, they can have no real measure of the potential market, for it may not be known precisely what is going to be marketed, where, at what price, under what competitive conditions. They can only assume the

Figure 7-2. The Project Life Cycle.

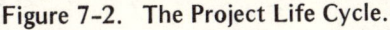

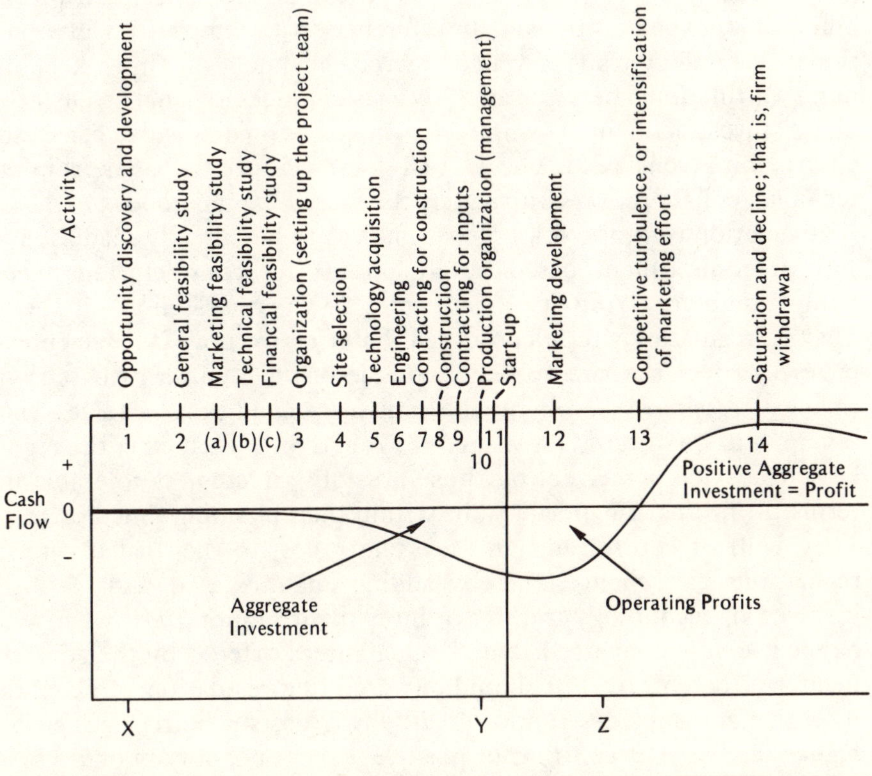

best and worst scenarios, as informed as they may be. *Risk* for the participants in an enterprise is greatest at the point of maximum investment (or uncompensated technology transfer), that is, at the point of maximum exposure prior to certainty of outcome (say, at point Y). By this time, enterprise participants should be able to estimate the probability of various financial results. Costs, prices, competition, consumer response can now be fairly accurately estimated. Both uncertainty and risk, in the sense we are using the terms here, evaporate to virtually zero at some point in the curve at which investment and a responsible profit have been captured (say, at Point Z).

Distinctions

This brings us to a major difference among the four policy levers, namely the timing of application during the project life cycle. Other differences relate to specificity of application, tax consequences, measurement of cost, visibility, risk of non-performance, and possibility of unexpected impact. In reference to the impact on international technology transfer via commercial channels, one must keep in mind distinctions between the six basic modes of such transfer— direct investment in a wholly or majority-owned subsidiary, an equity joint venture, a contractual joint venture, a partnership, a technical collaboration contract, or the sale of capital goods.

Regulation cannot work unless some entity has already made some sort of commitment based purely on market considerations, certainly not before Activity 4 in Figure 7-2 (site selection), if then. That is, regulating site selection will be effective only if some enterprise perceives the site to be acceptable on its own terms based on financial feasibility, even though it is a "second-best" choice. The same could be said of Activities 5, 6, 7, and 8. In each case, the regulation must be perceived, by those possibly affected, as tolerable in terms of internal financial results within their planning time-horizon. They will, of course, add an uncertainty cost to their calculations; regulations can change, thereby adding unanticipated costs. Those involved in technology transfer or investment cannot *know* what will happen, *unless* convinced that the guarantees offered by the government will be enforced. It should be noted that regulation—unlike tax incentive and subsidy—cannot create lower costs, a better market, or higher returns. Hence, it is not positive, except possibly in one case—

a regulation prohibiting additional entry into a market, thereby creating the possibility of monopoly profit for the enterprise. In any event, my own research in U.S. corporations[2] indicates that few managements, at least in the United States, are willing to make substantial investment or to transfer technology, on the assumption that a regulation will be enforced effectively or will continue indefinitely. There are too many cases to the contrary on both counts.

An uncertainty-reducing guarantee cannot be realized by its recipient until after a technology has been transferred or production started up, possibly only after revenues have begun to flow. I have found that an *up-front government commitment to purchase* a certain percentage of an enterprise's output at a specified price for a given time can be singularly effective for U.S. managers, although management must perceive the existence of an attractive share of an ongoing competitive market.[3]

Managements, I have learned, are somewhat more skeptical of—and, hence, less influenced by—government promises to prohibit entry of additional firms into a market. Historically, too many governments have reneged, as eventually they must—or install some sort of price control to prevent extraction of monopoly profit. In any event, the real implementation of such a promise comes well after production start-up. However, commitments to render breach of contract and expropriation of assets actionable under international law are possibly important to a management right from the start of significant investment or technology transfer, say between Activities 3 and 6 in Figure 7-2 (organization, site selection, technology acquisition, and engineering).

As already indicated, a tax incentive can be applied by a government only after some commercial activity has been commenced insofar as the local enterprise is concerned. Of course, a tax incentive for a foreign technology supplier, such as a reduction in withholding tax on royalties, can become effective as soon as any payments become due. If sales or value-added taxes are important to the recipient of technology, their reduction can impact as soon as Activities 5 and 6 (construction and contracting for inputs) begins. The same is true of a tax reduction on imported capital equipment and materials, that is, lowered customs duties. The reduction of taxes on capital equipment is of special importance in that it comes at the time of maximum risk and means a one-time reduction in costs which cannot be taken away subsequently. The enterprise management can see immediately whether the tax reduction is, in fact, implemented. If not, it could still withdraw with minimal loss in many cases.

Tax incentives promised in anticipation of a flow of profits (that is, a reduction of profit taxes) are rarely effective in stimulating the development of new projects. It would appear that such incentives are seldom factored into investment or technology transfer decisions. Uncertainty and risk of the venture's ultimate profitability are the commanding factors, so likewise uncertainty as to whether the government will live up to its commitment.

In contrast, a subsidy may be applied effectively for any activity from opportunity development to saturation and decline, that is, firm withdrawal.[4] Whether dealing with technology transfer via contract or direct investment, no purchase of inputs or commercial activity needs to have taken place. This means that subsidies can be given during those activities identified with periods of highest uncertainty (opportunity development—for example, for a new export—feasibility studies, and organization), for which neither regulation, tax incentive, nor uncertainty-reducing guarantees are really applicable.

Unlike a regulation or tax incentive, a subsidy or guarantee may be given quite easily by a government to specific, named enterprises. The point is that under many bilateral friendships, commerce and navigation treaties, trade treaties, and tax treaties, the signatory nations pledge themselves to desist from imposing discriminatory taxes, laws, and regulations. Subsidies and guarantees normally are not mentioned in such treaties, other than export subsidies and market protection which are, for example, restricted by the General Agreement on Tariffs and Trade. A country may determine, however, that it wishes to attract investment and/or technology, or technological and managerial input, from specific corporations or countries. It may do so either because it "knows" certain companies and has confidence in their sense of responsibility and technical proficiency, or because it wishes to diversify national sources of investment and technology, or because the country desires (or is obligated) to give preference to certain countries with which it has special relationships (as in a bilateral trade agreement, a free trade area, or a common market). In such cases, a host government may be able to offer appropriate subsidies and/or guarantees without running afoul of its international treaty obligations, which it might do if it were to apply discriminatory tax or regulation. Another distinguishing feature among the three approaches (regulation, tax incentive, and subsidy) is the tax consequence. A regulation or guarantee generally has no direct tax consequence, but tax incentives and subsidies are another story.

The tax liability of a firm to its parent government on income from foreign investment (or from a transfer of technology, technical or managerial skills), may be increased by reduction of taxes on local inputs (such as sales, value-added, or transactions taxes) or by subsidy. When earnings are remitted to the parent corporation, its taxable income at home may be increased by the amount of the foreign tax reduction or subsidy. That is, if a firm's parent country were taxing foreign-source income when remitted, as though it were domestic income (as does the United States and a number of other countries), something like a third to a half of the added profit (depending upon the income tax level in the investor's or technology supplier's parent country) would end up as tax revenue for the foreign—that is, the investor's or supplier's parent—government. Such taxes give rise to no tax credits which may be applied to the firm's parent country tax bill. They are "expensed," that is, they are deducted from taxable income as other business expenses in that they do not qualify as income or wealth taxes.

A tax imposed on profit by a host government may, of course, normally be credited dollar for dollar against the parent government tax bill, provided the latter taxes worldwide income. If the host government tax were reduced or waived, no benefit would accrue to the investing firm unless it were to reinvest its profit abroad—or unless the parent government either recognizes the tax-sparing principle (which the United States does not), or unless the parent government does not tax foreign-source income (or taxes it at a substantially reduced rate). In the worst case, the investing firm benefits from a reduced or waived profit tax abroad only to the extent of the interest on the amount it otherwise would have had to pay in tax to its parent government for the period in which the relevant earnings are held abroad. The appendix to this chapte provides numerical examples of the tax consequences of a subsidy, complete and partial reduction of taxes on inputs, and reduction of the profit tax on a firm. It should be noted that under U.S. law, a corporation has the option of either expensing or crediting foreign profit taxes, but under most circumstances crediting is preferred.

A government cannot quantitatively measure the cost or benefit of a regulation or guarantee, with the possible exception of government procurement. A government can only assume what would have happened if there had been no regulation or guarantee. What investment or activity has been foregone? How much does acquiescence to the regulation cost the firm involved in out-of-pocket expenses (filling in forms, delay, expert assistance) and in foregone opportunities? To what extent will the costs imposed by regulation be passed on to the consuming public via heightened prices or to the government in the form of lower taxes (by reason of reduced profit and/or purchase of inputs)? And what benefit will a guarantee ultimately bring to the firm, unless it be in the form of government purchase?

The cost and benefit of income or profit tax incentives are likewise difficult to calculate. In the absence of such incentives, what would have happened? The cost of such incentives to the host government includes the foregone tax revenue. But was the tax critical to the relevant enterprise decision? According to my research among U.S. corporations,[5] very few overseas opportunities are either abandoned or implemented because tax incentives are absent or present. Managers tend to view tax incentives as "frosting on the cake," which could very easily melt away if the host government were to change its mind.

The situation is quite different for subsidies for two reasons. First, the direct cost of a subsidy is known. Secondly, its effectiveness can be tested. A host government need not offer a subsidy at the start of discussion of a project, but only when it is perceived that a subsidy might be critical—and justified because of externalities. Many international executives cite subsidies as peculiarly effective because subsidies are for specified amounts, or their cash value can be determined fairly accurately and are paid early on ("up-front") at the time of high uncertainty and high risk for the parties involved—prior to either the transfer of technology or commercial production (in the case of direct investment). Such subsidies are generally plugged into financial feasibility studies. As indicated, management response to tax incentives tends to be very different.[6]

Perhaps equally effective is the promise of reduced taxes on the acquisition of technology, the procurement of capital equipment, and on the purchase of land and buildings. Again, the promise is timed so that it corresponds to a firm's period of high exposure.

Reduction of such taxes not only reduces the amount of exposure, but their "up-front," one-time nature reduces the firm's perceived uncertainty of actually receiving such tax incentives. Also, the impact of these reduced taxes on production costs can be calculated quite easily. And, inasmuch as such benefits bestowed by a government on an enterprise are not dependent upon long-delayed financial results, the cost of the benefits to the government can be more precisely calculated than can the cost of a reduced tax on profits. The cost of reduced taxes cannot be measured with the same precision as outright subsidies, but almost so if the engineering has been completed. Of course, the earlier in the development of a project such incentives are promised, the less exactly can the host government measure their cost, whereas a "true subsidy" (that is, a grant or specific payment for performance) is almost always subject to precise measurement, although the need for the subsidy may remain somewhat problematical. Was it really necessary to induce the enterprise to do whatever was intended?

Although subsidies thus possess several advantages over regulation, tax incentives, and guarantees, they have two important weaknesses—greater visibility and greater risk of non-performance.

Just as subsidies may be awarded more easily on a disciminatory basis (to specific firms, projects, or nationalities) and are subject to more precise measurement, so likewise are they more visible. A subsidy of $1 million given to Corporation X—either in grants of land or buildings or in outright payments to secure certain performance (such as the acquisition of a foreign technology)—is highly visible. In contrast, a tax incentive in the form of reduced tax on profits or on purchase of inputs is much less visible. The amount of the incentive is not specified, only a percentage reduction after some commercial activity has taken place (with the exception of a tax reduction on fees and royalties, or on the import of capital goods, paid prior to commercial production). In any event, the local public attitude toward the payment (or non-payment) of taxes may be very different than that toward the giveaway of visible public resources. The reward, in the former case, comes after the firm has shouldered major risk; a subsidy may not. Subsidies thus tend to become politically vulnerable, particularly if the recipient is a large, foreign-owned corporation and the donor government has what some perceive to be unfulfilled obligations to its own citizens.

For the very reason that preproduction or pretransfer ("up-front") subsidies are apparently so effective in terms of attracting management attention, they may be high-risk from the point of view of the local parent government. In that such subsidies are given before the event to induce an enterprise to develop a project, undertake feasibility studies, transfer or acquire technology, set up a project team, select a site, do the relevant engineering, etc., there is little to prevent a corporation—particularly a foreign corporation—from taking a subsidy and not following through, other than its sense of integrity and fair play. The firm may, at that point, have assumed little risk in terms of exposure. Hence, where an up-front subsidy is given it may make sense for a government to require the posting of some sort of performance bond by the firm, although this is not the general practice and would largely negate the appeal of a subsidy.

Of course, a host government's promise of a subsidy, payment of which is *contingent* upon the firm's doing something (for instance, effective transfer of technology, employing a given number of local personnel, undertaking local research and development, etc.), is somewhat less risky for the government. However, by the same token, it is less effective, because there is always doubt in management's mind that the government will, in fact, live up to its commitment. On the other hand, an uncertainty-reducing guarantee involves little or no risk to the government because it is triggered only upon some specified event within the government's control; but again, management may doubt compliance, *unless* underwritten by an internationally enforceable commitment.

It is probably true that a subsidy is more likely to achieve a desired impact without a counter-productive secondary effect. For example, a payroll tax to finance the training of skilled labor used by the firm adds to the firm's labor cost, thereby inducing it to substitute capital for labor. Likewise, the elimination of import duties on capital equipment, out of a desire to promote employment by increased investment in productive capacity, may in fact increase the capital intensity of a project by rewarding the use of capital (that is, reduce its cost to the firm), but not of labor. It might be more effective to pay an outright grant for each man-year of employment offered by a project.

Concessionary prices—another form of subsidy—may also produce unintended effects. An example would be a concessionary interest rate which encourages local borrowing, but only for fixed assets in that such rates are almost never available for working capital needs. Consequently, the more capital-intensive project is rewarded, even where the concessionary rates were offered presumably to stimulate employment by encouraging expansion of productive capacity. The extreme case of concessionary prices is where land and/or buildings are provided on a grant basis, or on a long-term, concessionary, leaseback basis, where the buildings have been constructed by the government specifically for the investing firm. The cost of capital for the firm is thereby reduced, and a more capital-intensive project than might otherwise have been the case is encouraged.

A subsidy that effectively reduces the price of an input—be it electric power, water, certain raw materials, land, capital—has the result of stimulating greater use of the subsidized factor, but by doing so, may cause a different technology to be employed. In the longer run, as subsidies are removed or reduced and the relevant inputs are priced closer to the market price, the subsidized technology (and associated skills) may prove to be uneconomic. The opportunity cost of the subsidies should be examined very carefully by both government and management regarding the costs induced when an enterprise employs other than the least costly mode of production.

Payment of a premium price for locally manufactured or processed products is another form of subsidy to the producing firm, but one so far removed from the original commitment to explore and develop a project that its cost-effectiveness may be limited. It may, however, stimulate technology acquisition by supplier firms. Of a somewhat different nature, the guarantee of government purchase of a specified minimum amount of the product at a reasonable price (reasonable in relation to a truly competitive price), over a specified period of time may, on the other hand, be cost-effective. Such procurement gets a firm past the difficult start-up period. So likewise does permission for a firm to import goods and technology tax-free for a period prior to or during the early phases of a project in order to test the market. I have found that this strategy is a little used, but very effective, device to stimulate interest in foreign firms to become involved in local production, whether via capital investment or technology transfer via contract.

104 / SPECIAL ISSUES

Table 7-1. Appendix: Financial Consequences of Tax Incentives and Subsidies When the Parent Government Taxes Worldwide Income.

	With No Incentive or Subsidy, Profit Tax Credited A	With No Incentive or Subsidy, All Taxes Expensed B	With Taxes on Inputs Reduced 50% Profit Tax Expensed C	With All Taxes on Inputs Waived D	With Profit Tax Waived E	With a Subsidy Equal to One-third of Costs F
1. Revenue	$100	$100	$100	$100	$100	$100
2. Deductible expenses						
(a) Taxes (non-profit)	(20)	(20)	(20)	(20)	(20)	(20)
(b) Other expenses	(60)	(60)	(60)	(60)	(60)	(60)
3. Taxes waived	0	0	10	20	0	0
4. Subsidy	0	0	0	0	0	20
5. Actual costs	80	80	70	60	80	60
6. Profit before foreign profit tax	20	20	30	40	20	40
7. Foreign profit tax (30%)	6	6	9	12	0	12
8. Remittance	14	14	21	28	20	28
9. Tentative tax liability to parent government (34%)	6.8[a]	4.76[b]	7.14[c]	13.6[d]	6.8[a]	13.6[d]
10. Foreign tax credit	6.0	0.00	0.0	12.0	0.0	12.0
11. Tax liability to parent government	0.8	4.76	7.14	1.6	6.8	1.6
12. Host government revenue[e]	26.0	26.00	19.0	12.0	20.0	12.0
13. Parent government revenue[f]	0.8	4.76	7.14	1.6	6.8	1.6
14. Total tax burden for firm less subsidy[g]	26.8	30.76	26.14	13.6	26.8	13.6

a. .34 × 20; b. .34 × 14; c. .34 × 21; d. .34 × 40; e. 2a + 7 − (3 + 4); f. 11; g. (12 + 13) − 4.

Assumptions:

Foreign government profit tax is 30 percent.

Domestic (parent) government profit tax is 34 percent.

Appendix: Financial Consequences of Tax
Incentives and Subsidies When the Parent
Government Taxes Worldwide Income

The six columns indicate the results in financial terms of six different policies in respect to tax incentives and subsidies. The actual results in each case will depend on: (1) the level of taxes imposed on inputs (sales, transactions, and import taxes), (2) the extent to which they are waived by the host government, (3) the level of profit tax imposed by the host government, (4) the level of profit tax imposed by the parent government (assumed to be 0.34 in this example), (5) whether the host-government profit tax is expensed or credited (assuming that the firm has this option), and (6) the level of subsidy provided by the host government.

As can be seen in our example, the most favorable cases for the enterprise are either the waiving of taxes on inputs (Column D) or the subsidy (Column F). Its total tax burden is $13.60 in each case. For the host government, the most favorable tax outcome is with no tax reduction (whether credited or expensed).

NOTES

1. Refers to the Coordinating Committee, the members of which represent the North Atlantic Treaty Organization countries, less Iceland and plus Japan. COCOM is charged with controlling the shipment of strategic goods and technology to Communist countries.
2. Richard D. Robinson, *Performance Requirements for Foreign Business: U.S. Management Response* (New York: Praeger, 1983), pp. 122–23.
3. *Ibid.*
4. Although I have no knowledge of any such subsidy in practice, theoretically a government might pay a firm to retire a product or entire venture so as to make it worth its while to do so (for example, a drug found to be injurious, after the market had been developed). A government might do so in order to reduce investor perceptions of costs arising out of uncertainty in respect to government intervention, and thereby encourage subsequent investors to enter the market.
5. *Ibid.*, pp. 117–120.
6. *Ibid.*, pp. 116–17.

CHAPTER 8

The Siting of Research and Development Abroad

Calculations as to total national expenditures for research and development, such as those charted in Figure 8-1, may be very misleading. Not only do they include military R&D expenditures, which are notoriously unproductive when it comes to commercially valuable technology, but they *exclude* the R&D which firms based in a given country are carrying on through affiliates elsewhere. These foreign affiliates may be wholly or majority-owned subsidiaries, but probably increasingly are joint enterprises of one form or another. (See Chapter 11.)

NEED, PRACTICE, JUSTIFICATION

Much that has been said in preceding pages emphasizes the need for a situationally oriented research and development effort, whether one speaks of product, process, or machines. A research and development division embedded in a domestic operation, that is, one in which foreign markets are only in the peripheral vision of management, is unlikely to develop the interest and skills necessary for devising optimum solutions to the problems posed elsewhere in this volume. Also, the cost of R&D may be lower in some regions than within the home country, though differentials may be narrowing. A compromise solution devised by some companies is the employment of technical liaison personnel to provide the linkage between foreign

Figure 8-1. Civilian Research and Development Expenditures, 1960-1984 (as a Percent of GNP).

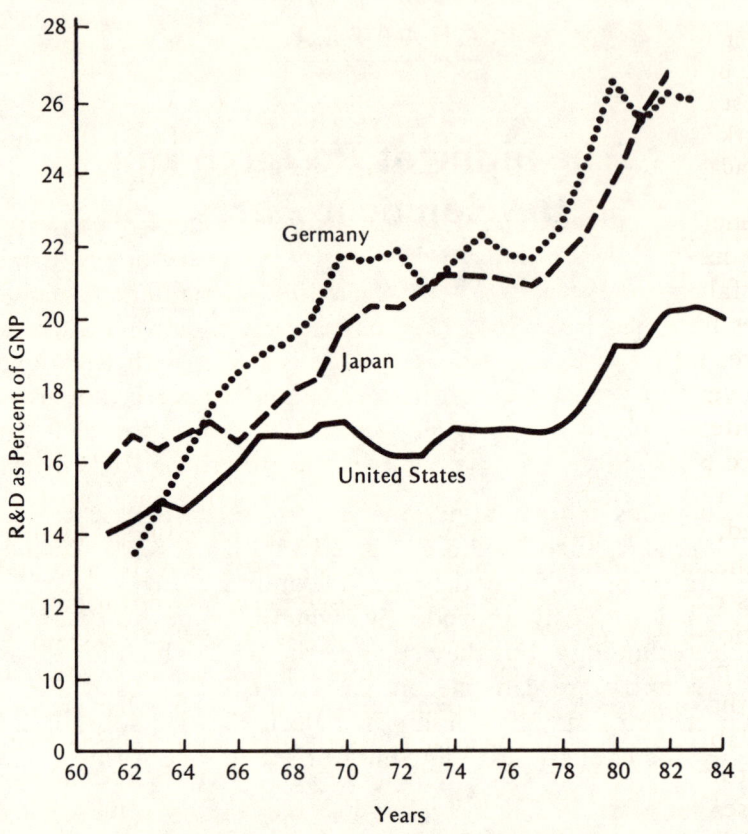

Source: National Science Foundation, taken from *American Excellence in a World Economy* (a report of the Business Roundtable on International Competitiveness, June 15, 1987, Washington, D.C.), p. 20.

market environments and domestically based R&D sections. This linkage, of course, may prove to be the transmission belt for a valuable two-way technical flow as a firm's competitors abroad or local foreign researchers come up with new ideas and products. This externally induced stimulation may be important in devising the optimum entry into a given market. There is, of course, an important feedback into sales strategy area; constant adjustment is needed. Organizational structure should be such as to facilitate this need.

A Conference Board study of U.S. business as early as 1970 observed:

> The research and development function is not one that can be easily decentralized. In spite of their desire to make maximum use of the capabilities of their foreign units, most companies cooperating in the survey make limited use of them for R&D. In spite of pressures to decentralize research activity, most companies carry out the bulk of it in the United States.... When R&D work is done overseas, it most often takes the form of product modification or adaptation to meet the particular needs of local markets.[1]

Generally, it seems there is an overwhelming concentration of R&D in the firm's home country. In recent years, as profit margins have fallen and international competition has tightened, many companies have had to review costs, particularly those of a discretionary nature. Often near the top of the list is R&D. With R&D expenditures very frequently ranging between 2 and 10 percent of gross sales for internationally competitive firms, it is natural for such firms to reduce their commitments to long-range growth via R&D investment when the immediate concern is for short-term survival. It is also argued, with some justification, that a centralized R&D facility serving the entire corporation is the least costly way to accomplish a firm's research objectives. Even if some decentralization of the R&D function had been planned, possibly even carried out, pressures seem to be pushing in the opposite direction. There can be no denying the fact that duplicate laboratory facilities are costly, that staffing foreign laboratories can lead to increased hiring and training costs, and that total cost control is made more difficult by the decentralization of research activity. The real issue, however, is cost-effectiveness, not simply cost control. Benefits as well as costs determine the optimal solution to the problem of research location.[2]

Control is closely tied to cost-effectiveness. Perhaps the most important element in a research program is the effective integration of all components of the effort. Obviously, integration is most readily accomplished when all aspects of the R&D program are under one roof, let alone scattered around the globe. A frequent criticism of decentralized research is the likelihood of redundant activities appearing in various laboratories. Researchers naturally wish to become involved in the most promising projects. This tendency can lead to the deliberate disguise of projects in order to counter charges of redundancy, or to flagrant duplication of effort. Should such situations exist, there is certain to be loss of cost-effectiveness.

Many, if not most, of the firms involved significantly in international business seem to take pride in their R&D. It can be expected that top management in such firms feels a need to be in close touch with this activity, either to make inputs into the programs or merely to remain informed of progress. But whatever the reason, this personal interest can be a decisive influence in locating R&D.

Another factor of relatively recent origin is the collapse of the dollar in relation to certain other currencies, thereby reducing the relative cost of U.S.-based R&D. It is admittedly risky, nonetheless, to locate R&D merely in response to shifts in foreign exchange rates, which may prove temporary.

Another point needs to be made in this context. International managers may find that the best opportunities for the growth of their respective firms are outside their home countries. A very strong case can be made for locating R&D activity within a firm's principal markets in order to improve the responsiveness of the company's technical programs to local needs. This responsiveness can be measured in both time and relevance. Too often the foreign-based company attempts to promote its domestically developed products overseas without considering their appropriateness to the market. Without an effective technical staff on the scene, it is very difficult to ascertain the peculiar product characteristics desired, and certainly it is most difficult to determine the needs in a timely manner. This can prevent the firm from either acquiring, or maintaining, a competitive edge.

Closely related to the above issue is the desirability of developing and maintaining a global product. It is often possible to design characteristics into a product that will satisfy multiple market areas. But for this to be done during the development stage, it is imperative that market needs from around the world be known and appreciated. Acquiring such knowledge is difficult without experienced technical personnel on hand to ascertain market needs and to funnel these through design requirements into the central laboratories. Such organization can result in R&D cost-saving since the need to redesign products may thus be minimized.

Another strong argument for locating technically trained staff in foreign locations is related to industry leadership. For example, the electrical power transmission industry is said to be more highly developed in Western Europe than in the United States. A U.S. firm's failure to participate in this leadership by not employing European R&D personnel could lead to competitive disadvantages.

Most R&D-oriented firms utilize their technical personnel to support sales and marketing efforts. Beyond the obvious use of technical service personnel, the deployment of research and product development personnel within a market area can lend credibility to a firm's sales program. In countries with indigenous competitive firms that maintain full R&D facilities, the foreign-based company relying totally on one-country R&D may find itself at a considerable sales disadvantage. Regionally based research activities may also be useful in keeping technical service personnel upt-to-date on the latest corporate technology. The technical service engineer who does not have the opportunity to work with his research and product development colleagues on a continuing basis may well become technically obsolete within two or three years. This obsolescence could reduce his effective product knowledge to that of a mere salesman, thereby wasting the money spent to maintain him in the area and reducing the competitive edge of the firm.

High-technology companies often develop management personnel from the ranks of the R&D departments. If the firm operating internationally wishes to follow a similar pattern overseas, a research department will be essential to any foreign subsidiary or affiliated enterprises. Continued reliance on more marketing-oriented management candidates can seriously alter the profile and competitive characteristics of a firm engaged in the application of high technology.

An important activity for a high-technology company is that of maintaining liaison with the academic community in the principal areas of company operations. This liaison is not possible without technically trained and active company personnel on the spot. The firm that fails to keep in touch with the main institutions around the world involved in basic research relevant to its industry is very likely to miss opportunities for growth and innovation. This same principle applies to active participation in professional societies and similar industry associations. The existence of a solid R&D function in the major areas of involvement is central to establishing these kinds of relationships.

A final argument often advanced for maintaining a research activity in selected countries where the firm does business concerns corporate citizenship. Many host countries demand a high level of participation by the foreign firm wishing to operate in that country. The absence of a company-sponsored R&D facility that can employ the technical expertise of the host country's universities, as well as adding to the technical base of that country, may be seen as something less than total involvement.

An optional strategy is for the firm to enter into a continuing supportive relationship with one or more of the relevant research institutes within a host country, either through R&D contracts, the loan of personnel, or by general financial support. These research institutes, often industry-specific and often partially supported by government funds, are a common feature of the contemporary landscape in both less developed and more developed countries.

The many hundreds of transnational research cooperation agreements being reported out of Europe suggest a certain economy to be achieved from increasing scale of an R&D effort. Because of the difficulty of assigning to the member firms commercially valuable discoveries and developments emerging out of these collaborations, there may be a tendency for *jointly owned* ventures to exploit the R&D. Over time, with repeated joint venturing of this kind, the parent firms will have, in effect, merged.

A publication of *Business International* observes:

> Companies are directing special efforts to develop research within Latin America. Johnson & Johnson reported that the research departments of its larger Latin American affiliates are among the most active of such organizations engaged in pharmaceutical development. Ralston Purina's Latin American research programs, staffed 100% by nationals, have yielded such a high level of expertise that two or three staff members are candidates to head the company's world-wide R&D program. The Argentine subsidiary of General Tire has even gone further in promoting R&D by teaming up with a local group to explore advanced rubber technology.[3]

But generally speaking, it seems quite clear that most firms operating internationally are exceedingly reluctant to put any significant R&D effort into their LDC operations. Among those factors discouraging the location of R&D in the LDCs can be cited: (1) import restrictions, in that they reduce competitive pressures; (2) relatively small size of many LDC markets; (3) shortage of trained personnel; (4) the ineffectiveness of local law regarding patents and protection of commercial secrets. However, "as research costs soar at home and standards of university education rise abroad . . . research work could well begin to move to less expensive areas close to the production site."[4] It remains to be seen how rapidly this movement will, in fact, occur, if at all.

A Soviet analyst, projecting present trends into the future, has estimated that by the end of the century the LDCs' bill for imported technology will be twenty to thirty-five times the present annual expenditure of $3–$5 billion. Noting that U.S. corporations operating internationally spend only about 4 percent of their total R&D funds abroad, he doubted that such corporations will do other than increase technological dependence. He advocated that technology transfer contracts (and direct investment projects) be required to cover the setting up of laboratories and of pilot training facilities, "so that the imported technology can serve as a threshold for further progress on the spot." He went on to point out, "We in the U.S.S.R. generally insist that R&D facilities be supplied with the production lines."[5] But what he did not say is that the Soviet market and consequent scale of production were far different from the typical LDC, and he himself warned against any country's shutting itself off from external technological development.

Another aspect of R&D should be noted: some countries offer subsidies, from outright grants to tax write-offs, if the activity satisfies certain criteria. A few national examples are cited below, though these programs are subject to frequent change. In Canada, to qualify for an R&D grant, a firm's expenditures must be for scientific research and development to strengthen or extend its business within the country. Corporations must therefore undertake to exploit the results of their research and development work in Canada unless, according to sound business judgment, it would be uneconomical to do so. Furthermore, enterprises must normally be free to market products resulting from their research and development to all countries of the world.[6]

In France, DATAR (Délegation à l'Aménagement du Territoroire et à l'Action Régionales, the government's regional and national development agency) offers direct grants up to 20 percent of an investment in R&D (depending upon its size and location) for fixed assets and the relocation and training of personnel. R&D centers in certain regions enjoy a number of tax relief measures. Additionally, various government agencies are authorized to enter into R&D contracts with private firms and to provide loans and credits for their execution.

Another country offering official inducement for R&D is Japan. The vehicle for assistance is the Research Development Corporation, the purpose of which is to stimulate the development of commercially valuable technology from research produced by universities, private research institutes, and private companies unable financially to carry out development. The corporation selects that research deemed to be of greatest developmental value and advertises for bids for development by private firms. A successful bidder then borrows the necessary funds from the corporation, which are to be repaid (within five years at no interest) if the development turns out successful. No repayment is required in case of failure. If a company thus commissioned exploits the technology commercially, it must offer a certain royalty rate to the corporation, which in turn divides it with the original owner of the technology.

Japan may be something of a special case. In late 1987 it was reported that nearly twenty U.S. and European chemical, pharmaceutical, and electronic firms had created research and development centers in Japan or had announced plans to do so in the near future. Their alleged motive was to acquire sophisticated applications technology through close contact with Japanese industry. Without doubt, commercial R&D is rapidly becoming a worldwide phenomenon.

In the 1970s, the European Economic Community (EEC) moved to strengthen intra-European Community technological cooperation through a new device, the Community Industrial Development Contract (CIDC). A CIDC is drawn up to fund a European Community-based joint effort of two or more companies headquartered in two or more EC member states, through low-interest loans administered by the European Investment Bank. A CIDC is granted for a project of sufficient financial and/or technological risk that it could not be realized without Community or national assistance. Some additional criteria used in evaluating proposals for a CIDC include: (1) contribution to the technological development of a specific industrial sector; (2) degree to which imports from non-EC countries could be reduced; (3) creation of new export opportunities; and (4) satisfying major economic or social needs of member countries. The original provisions of the CIDC gave high priority to certain advanced-technology industrial sectors, including pharmaceuticals. Subsequently, however, these sectors were removed from priority status in order to provide equal access to a CIDC for a range of sectors. Presumably, subsidiaries of non-EC firms are not discriminated against. Industrial property rights arising out of a CIDC belong to the companies involved, subject to the general provision of the EC patent.

In a 1978 study of government influence on the process of innovation in Europe and Japan, it was found that such influence had been "felt in one way or another on about 43 percent of the 164 projects in the sample."[7] The report observed:

> Governments are able to control the inputs to R&D projects and thereby influence the rate and direction of technological change through their ability to alter the industrial environment—market forces, industrial organization, the structure of rewards, and the regulatory constraints within which industry operates. Governments can also alter the resources available to firms—information, financial and human resources—and the allocation of these resources. In affecting either the environment of or resources available to industrial firms, governments can help to initiate technological change, to sustain change, or to regulate change in such a way as to ameliorate its effects. Since the number of possible modes of action available to government to influence technological innovations is large, it is necessary to use some sort of simplifying concept to facilitate further analysis.[8]

The simplifying concept used was to specify twelve mechanisms employed by governments, the first seven being mechanisms affecting the innovation process itself (initiating), three mechanisms affecting resources (sustaining), and the final two mechanisms to ameliorate the consequences of technological change (restructuring). The twelve mechanisms:

1. Stimulating innovation by working through market forces
2. Reducing the cost of innocative activities
3. Reducing the probability of technical or commercial failure
4. Increasing the rewards to the firm for successful innovation
5. Encouraging innovation via market entry by new firms (or existing firms in new markets)
6. Restructuring an industrial sector (by encouraging mergers or acquisition)
7. Influencing the organization and management of individual firms (for example, by direct communication and persuasion)
8. Influencing the availability, utilization, and mobility of managerial and technical manpower (for example, by training, allocation, and subsidy)

9. Assisting institutions (universities, research institutes, private consulting firms, industries, and governments) with regard to the generation and utilization of technical knowledge
10. Increasing the transmission and transfer of technical knowledge between institutions (for example, by conferences, publications, exchange of people)
11. Ameliorating the adverse consequences of technology with respect to the environment, natural resource utilization, and public safety
12. Influencing labor's receptivity to technological change and internalizing the human costs associated with innovation activity (possibly by government subsidy such as retraining and relocation)

It was found that 57.2 percent of all the projects studied had been influenced positively or negatively by government involvement via initiating mechanisms; 37.2 percent via sustaining mechanisms; and 27.2 percent by means of restructuring mechanisms. However, there were substantial differences by industrial sector. Likewise, considerable differences among countries were found in respect to the relative dependence on the various forms of government involvement.[9]

Previous, possibly now out-of-date research indicated that perhaps not more than 10 percent of all R&D expenditures by firms with international operations has been directed into overseas R&D activity.[10] Perhaps one could double this percentage to 20 percent today, but no one seems really to know. In any event, only about one-third of that total, whatever it may be, has been directed to the less developed countries, or between 6 and 7 percent of the total R&D expenditure by these firms.[11]

Behrman and Fischer, writing in 1980, concluded:

> The data presented indicated that a considerable amount of R&D is presently being performed abroad by both American and European transnational corporations. While most of this R&D is located in developed countries, there is diversity among the locations selected. This diversity, in part, reflects the varying market orientations of the firms. Market orientation also appears to be an important determinant of the means by which a transnational corporation [read, corporation with international activities] establishes R&D abroad.
>
> The evidence presented suggests that . . . corporations with "host market" orientations are generally most likely to pursue R&D abroad, particularly when developing countries are at issue. They are, however, less likely than

their "world market" counterparts to delegate new product research responsibilities to their foreign laboratories, and they also rely heavily upon their evolution from technical services as a means of establishing foreign R&D groups.[12]

The Japanese are also getting into the R&D act. In late 1986, it was reported that Japanese centers for R&D were being created "one after the other" in the United States and other technologically advanced nations. The avowed aims of these centers were the collection of the most up-to-date information on technology and production, the mastery of technical skills, and the provision of technical backup for local manufacturing facilities. The centers also represented "Japanese corporate counter-measures to growing moves in the U.S. to clamp down on the easy outflow of information on new technology." It was admitted by a Japanese source, however, that "the presence of industrial espionage still demands a legal framework that imposes limits on how far research is allowed to go."[13] The difference between an overseas "R&D" unit and an industrial espionage center (the "listening post" notion) may be a fine one.

The criteria relevant to the international siting of R&D appearing in the same source quoted above is worth replicating here. (See Figure 8-2.)

According to a study by Ronstadt, it appears that R&D activity of firms operating internationally tends to be institutionalized in four quite distinct ways: (1) the technology transfer unit, (2) the indigenous technology unit, (3) the global technology unit, and (4) the corporate technology unit. Conditions leading to the establishment of each type, as well as experience, is of interest.

Technology transfer units are established by a firm to help certain foreign subsidiaries transfer manufacturing technology from their respective parents and to provide technical services for local customers. Drawing on the Ronstadt study, one notes a common three-stage progression in establishing such units. First, the parent decides to make an investment in production facilities abroad. Second, foreign managers in charge of production and marketing activities discover that product and process technology is flexible and introduce minor modification and adaptation. Third, foreign managers see the value of an ongoing stream of technical service on behalf of customers.[14]

In another study,[15] it was found that each of the firms studied (ten U.S. and one Dutch-British) had established technology transfer units to effect minor product and process modifications, as well

Figure 8-2. Important Criteria for Considering or Not Considering Overseas R&D Locations.

	Home Market Firms	Host Market Firms	Worldwide Market Firms
Important criteria for considering an overseas R&D location	1. Proximity to operations 2. Availability of universities	1. Proximity to markets 2. Concept of overseas operations as full-scale business entities	1. Availability of pockets of skills in particular technical areas 2. Access to foreign scientific and technical communities 3. Availability of adequate infrastructure and universities
Important criteria for not considering overseas R&D locations	1. Products sold in the developing countries are not sophisticated 2. Lack of qualified scientists and engineers 3. Economics of centralized R&D	1. Increasing costs of doing R&D overseas 2. Economics of centralized R&D	1. Economics of centralized R&D 2. Difficulties in assembling R&D teams

Note: "Home market firms" are primarily concerned with investing abroad for the purpose of serving their domestic markets through the import of product. "Host market firms" are largely oriented to the markets wherein they are located. "World market firms" are those whose foreign affiliates are integrated to a degree to serve a standardized world market.

Source: Jack N. Behrman and William A. Fischer, "Transnational Corporations: Market Orientations and R&D Abroad," *Columbia Journal of World Business*, Fall 1980, p. 59.

as provide technical service for foreign customers. Most of these units had subsequently expanded their functions to include indigenous technology, but some had retained their original functions.

Examples developed in this latter study are illustrative. Corning Glass Works had set up three technology transfer units abroad which had retained their original functions without change, their function being to help apply electronic technology supplied by the U.S. parent. By way of contrast, Union Carbide had established technology transfer units for performing manufacturing and customer technical assistance service, which later evolved into indigenous technology units. The same development was reported in respect to technology transfer units set up by the Exxon Corporation. The original functions of these units had been technical assistance and product modification, but subsequently, their activities were expanded to indigenous technology development. The same phenomenon was reported in respect to the technology transfer units of six other of the eleven firms. In the case of IBM, however, the technology transfer units evolved into global technology units due to a different international marketing strategy discussed later in this chapter.

Indigenous technology units have as their major function the development of new and improved products expressly for foreign markets, products that are not the direct result of technology supplied by the parent organization.

According to the work referred to earlier, the four conditions resulting in the creation of an indigenous technology unit are: (1) substantial direct foreign investment in production facilities, (2) the perception that the future stream of new and improved products developed in the parent country will not satisfy the growth needs of the foreign business, (3) the identification by foreign managers of investment opportunities abroad which differ from existing domestic business, and (4) a decision that a major shift in R&D policy is necessary to cope with a specific investment opportunity abroad.[16]

Basically it would seem that companies set up indigenous technology units abroad as a result of two strategies: *first*, the creation of an indigenous technology unit directly; the *second*, enlarging the function of an overseas technology transfer unit so as to include the development and improvement of new products and processes expressly for the foreign markets.

The first strategy, for example, was adopted by Corning Glass, Otis Elevator, and Johnson & Johnson, three of the companies examined in Woo's 1983 study.[17] The first named had established two indigenous technology units, one in Belgium, the other in the United Kingdom. The former—Corning's European Research Laboratories—expended 50 percent of its effort in support of existing business activities, 20 percent in exploratory research in existing business areas, and 30 percent in the development of new high-risk business. The other unit, located in Halstead, U.K., had the objective of developing new medical products for the local market. The indigenous technology unit of Otis Elevator Company was mainly for developing new products and processes for Europe. It had begun to do so with the development of specially designed small escalators and elevators. The Canadian laboratory of Johnson & Johnson had begun some new product development and was trying to find a gap it could fill in the U.S. programs. For example, due to the small market size for a non-woven fabric in Canada, the Canadian laboratory had developed a new product with different manufacturing processes, applicapable to small-scale operations.

The second strategy, the staged development of indigenous technology units, the Woo study found quite common. This approach tended to expand the functions and activities of the technology transfer units from simply performing manufacturing and customer technical assistance services to the development of new and improved products and processes for the foreign market. Union Carbide, Exxon Chemical, Exxon Corporation, CPC International, Otis Elevator, du Pont, and Johnson & Johnson had all pursued this approach extensively at the time of the study (1983). The technology transfer units of Exxon Corporation, Exxon Chemical, CPC International, and Otis Elevator that had been created originally to perform primarily technology transfer work, had evolved by 1983 to the point that their R&D activities were geared *primarily* toward the development of new or improved products and processes expressly for the European market. They had also shifted into exploratory research for existing business activities.[18]

The market-demand pull operating in this evolutionary process could be demonstrated through du Pont's Belgian affiliate, whose sole original purposes had been quality control and technical service. However, because European customers would not readily accept the U.S. product line, product adaptations had been required for local uses. As it grew, the Belgian laboratory had responded to the pull of market demand by moving from the development of finishes (primarily oriented to color-matching and customer technical service) to the developmental stage in paints, and finally into applied research for new product adaptations. A similar market-demand pull had occurred in the British market of Johnson & Johnson. In one case, the British laboratory had picked up a "failure" in the United States which it felt suitable for the British market. In another case, the materials necessary to make a particular process work were available in the United Kingdom but not in the United States and the British affiliate picked up the idea and pursued it to fruition.[19]

It would appear that a local R&D activity, under the pressure of market-generated need, might respond independently of R&D activity elsewhere in the company. Since national markets are not identical, firms operating internationally are under considerable pressure to set up individual R&D units to develop new products to meet special needs and requirements.

Global technology units are created basically to develop new products and processes for a virtually simultaneous application in major markets, both foreign and domestic. Such a move implies that a corporation's strategy is to develop a single product line for worldwide distribution and that the R&D activities of different units are highly coordinated and controlled by the center. A necessary, though not sufficient, condition for such development is the existence of substantial production and marketing resources located abroad with which to associate R&D operations.

IBM is cited as a typical example of the establishment of global technology units. IBM's first foreign R&D units were created in England and France during the 1930s, largely for product modification reasons.[20] Later, development laboratories were started up in other countries, such as Canada and Japan, as a part of corporate-controlled development activity. This meant that these laboratories were international and corporate in nature rather than devoted to promoting exclusively the interests of a region, a national company, or a specific unit. Major R&D missions were assigned to the laboratories by IBM corporate headquarters. By 1974, each foreign developmental laboratory had worldwide responsibility for specific product-technology areas. The primary force stimulating the establishment of these development laboratories was an explosion of new product needs by computer customers, coupled with the growing competitive need to improve existing data processing products and processes in the United States, Europe, Canada, and Japan.

Suffice it to say that the creation of global technology units is probably limited to certain industries in that the products involved must be of such a nature that they can be introduced in different markets virtually without change. The number of such products is possibly quite limited.

Corporate technology units constitute the fourth mode of institutionalizing R&D. Their purpose is to generate new technology from different countries expressly for the corporate parent. Two factors may be involved in these units: first, the presence of top scientists abroad who cannot (or will not) be transferred to the parent corporation, and second, lower cost of foreign R&D. But the success rate of such units seems low.

For example, CPC International had set up corporate technology units in Italy and Japan.[21] The original objective of these units had been to perform R&D projects expressly for the U.S. parent by utilizing outstanding scientific talent found in these countries. However, the performance of these units had become unsatisfactory, so much so that the corporate technology unit in Italy had been disbanded. CPC International managers felt that the unit had not achieved any noteworthy successes. Moreover, the Japanese corporate technology unit had shifted into the generation of new products and processes for CPC's Far East affiliates. By 1976, it was forecast that the unit's activities would be soon split 50-50 between projects performed primarily for the U.S. parent and those for the regional affiliates.[22] Union Carbide in 1969 abandoned a corporate technology unit in Belgium, apparently because of lack of defined goals for research and of linkage between the R&D unit and marketing.[23]

Conversely, Eastman Kodak's corporate technology unit in Australia allegedly had made important contributions to the products and technology of the company.[24] This positive result was reportedly due to frequent visits of U.S.-based laboratory personnel and exchange of reports between the unit and the parent company.

This experience with corporate technology units, limited though it be, would seem to indicate that their success is based on the existence of excellent communication between the parent company and the R&D units, which seems to imply a considerable expense in order to utilize the expertise of foreign scientists without moving them.

One student of the subject concluded that firms operating internationally are very likely to concentrate on indigenous technology and global technology units in the development of overseas R&D. The indigenous units are designed to maintain close contact with foreign markets in order to develop new and improved products and processes for market-specific needs. Even though the cost of communication among the R&D units and the central R&D development of a firm may be expensive, it may well be justified in terms of improvement in the linkage of the R&D functions with the market. Nonetheless, this analyst went on to say that he would expect that the emphasis of R&D units abroad would be on the improvement and modification of products and processes originating in the parent company rather than on entirely new processes or products.[25] There is some evidence to support this conclusion in that the percentage of overseas R&D spending for *basic* research is reported to be about six percentage points less than the percentage of domestic R&D spending for such research, and the percentage of overseas R&D of an applied nature is ten percentage points less than in the domestic case. However, the percentage of overseas R&D spending for *development* is reported to be sixteen percentage points greater than the percentage of domestic R&D spending.[26] Hence, most overseas R&D seems to be focused on product and process modification and improvement.

In respect to the establishment of global technology units, greater worldwide integration of overseas and domestic R&D can probably be expected. In one survey of ten U.S. companies with international operations, almost half had accomplished worldwide integration of overseas and domestic R&D activities, and most of the rest indicated some limited integration had been attempted.[27]

R&D COST

It has been argued that the establishment of R&D abroad is, at least in part, due to its relatively lower cost. Table 8-1 reports the mean ratio of the cost of R&D inputs in Europe, Japan, and Canada compared with those in the United States. One notes a very substantial cost differential in 1965, although it had narrowed by 1975. Moreover, the ratio did not include the costs of communication and coordination implicit in maintaining a multinational network of laboratories. More recent estimates do not seem to be available. One can conclude, however, that the cost differential is no longer a determining factor in siting overseas R&D. This shrinking cost differential also explains the fact that reportedly the corporations have been trying to move their corporate technology units in the direction better characterized as indigenous or global technology units.

Table 8-1. Mean Ratio of the Cost of R&D Units in Europe, Japan, and Canada to That in the United States.

Year	United States	Europe	Japan	Canada
1965	1.00	0.68	0.56	0.82
1970	1.00	0.74	0.60	0.86
1975	1.00	0.93	0.90	0.96

Note: There are many costs of communication and coordination in a multinational network of laboratories, which costs are not included here. Data was based on a 35-firm sample, from which usable data was obtained from 19 firms. Many of the rest had no overseas R&D experience. There are considerable differences within Europe in the level of R&D costs. According to a number of firms in this sample, costs tended to be relatively low in the United Kingdom and relatively high in West Germany.
Source: Adapted from Edwin Mansfield, David Teece, and Anthony Romeo, "Overseas Research and Development by U.S.-Based Firms," *Economica* (May 1979): 192.

R&D IN LESS DEVELOPED COUNTRIES

There is evidence, already noted, that about a third of the total R&D dollars spent in overseas laboratories by the large, internationally active companies is in the developing countries and that this proportion is on the increase.

Even though there are some similar characteristics between the evolutionary process relative to overseas R&D activity in the less developed countries and that in the more developed, the overseas laboratories in the LDCs tend to be more concerned with developing ways of utilizing local inputs. It would appear that the general LDC case is an evolution from technology transfer unit to the indigenous technology unit.

For example, the R&D laboratory in du Pont's Brazilian affiliate was 60 percent dedicated to quality control work. Subsequently it began to expand its activities into improvement and substitution of local materials and the quality control work was shifted to the manufacturing plant. Moreover, anticipating market change, the laboratory was eventually reoriented toward various marketing demands and responded with new formulations and processes.[28] One of the objectives of Union Carbide's New Delhi laboratory was to provide its Indian subsidiary with the capability of developing an optimum process technology for India, given the country's resources, domestic labor costs, and new technological developments. Other objectives were likewise present, specifically the formulation and development of products for the Indian market which could be achieved more efficiently in India due to local cultural and economic conditions.[29]

As indicated, much of the overseas R&D in the developing countries is concerned with increasing local value-added by increasing local input. By so doing, a firm can establish a closer bond with the local economy, and the foreign operation is not as likely to be viewed as purely "foreign." For example, Johnson & Johnson's Brazilian R&D unit had tried to determine how to manufacture its products from local inputs. In doing so, it contributed to the diffusion of technical skills to local suppliers in the effort to improve product quality. This effort included providing suppliers with manufacturing specifications and assisting them to learn how to maintain quality standards and production schedules. The R&D unit also aided suppliers in buying specialized equipment which would produce to the specifications and volumes required.[30] The Indian laboratory of Unilever, it was reported, was involved in developing processes to upgrade indigenous raw materials as part of an import-substitution program and to improve the product line. The laboratory had developed a host-market orientation by evolving with the growth of the market and developing products not standardized with the parent. This laboratory was operating virtually without parent control.[31]

The development of these laboratories presumably saved considerable foreign exchange for the host countries through the substitution of local materials for foreign, encouraged scientific and technological employment, increased employment in supplier industries, and reduced the costs of production.

HOST-GOVERNMENT POLICY

Host-government policy can exert considerable pressure on the foreign-based corporations to establish local R&D units. One survey revealed that nineteen out of sixty-two foreign R&D laboratories were identified as having their origins in host-government pressure, three of these being joint ventures. The countries most active in attempting to influence foreign corporations in this respect were said to be Canada, Brazil, France, India, and Japan.[32] (Canadian, French, and Japanese efforts were discussed earlier.)

However, a survey of LDC government policymakers showed that only 7.7 percent considered that foreign corporations conducting

R&D directly in local problem areas were making an important contribution. They reportedly felt that contributions of foreign firms other than local R&D siting were more important.[33]

On the one hand, foreign corporations tend to consider local market potential to be more important than the host-government intervention in locating foreign R&D. On the other hand, some consider government intervention to be more important than the condition of the local scientific and technical infrastructure.[34]

THE RELEVANCE OF CORPORATE STRUCTURE

Structure may be an important factor related to whether or not a corporation has overseas R&D activities, and the type of R&D activities undertaken. A *Business International* study revealed that firms with a strongly centralized management were probably best off with minimal overseas R&D activity, unless some irresistible benefits existed, such as gaining the special expertise of foreign institutions or individuals. On the contrary, highly decentralized corporations tended to be more liberal toward R&D management, to establish more R&D units overseas, and to allow operating responsibility for R&D to fall to subsidiary or affiliate management control rather than corporate.[35] One would expect strongly centralized and more home-market-oriented corporations to be less involved in overseas R&D activities than less centralized, host-market–oriented corporations. This logic coincides with the idea that the evolution of indigenous technology units is characteristic of those corporations which prefer to get as close to the markets being served as possible and, hence, develop market-specific products and processes.

Although both centralized and decentralized corporations may possess central corporate R&D laboratories, such a laboratory, as operated by the highly centralized corporation, acts more as the leader of the firm's R&D activities and serves as the focus of the R&D activities of the corporation's foreign affiliates or partners. In contrast, in the decentralized corporation case, the central corporate R&D laboratory is likely to act in an advisory role to decentralized R&D units and tends to perform research of a long-range or new-venture nature which would not usually be pursued by operating divisions.

In general, the highly centralized corporations tend to introduce delays and rigidity into the project review process in that candidate projects are required to pass through several levels for approval. The R&D activities of such corporations, however, are likely to be more coordinated. On the other hand, the decentralized corporations, which allow a greater amount of autonomy to their overseas R&D units, may encounter cost in the duplication of R&D at various sites, a problem noted before.

Past and contemporary research indicates that the development of R&D activities in the more industrialized countries by foreign corporations is likely to continue largely in the form of indigenous technology units and global technology units. The basic functions of these overseas R&D activities are primarily product adaptation and modification for the local market, or for multiple markets in the case of a global technology unit. New product and process development is not common in R&D units located abroad in the industrialized countries because of the similarity of environmental factors such as levels of economic, technological, and human-resource development. However, one suspects that the link with local markets and local inputs is likely to be emphasized increasingly in future overseas R&D activity. As in the case of du Pont's Belgium laboratory, modifications are frequently required to permit substitution of local materials which have different qualities and reliabilities.

On the other hand, overseas R&D activities in the less developed countries are likely to concentrate more on the development of new products and processes appropriate for local markets. This expectation arises out of the differences in environmental factors characteristic of the LDCs, plus governmental intervention to encourage local R&D.

As indicated, cost differentials of R&D activity in different countries are not likely to be a determining factor in the decision to site R&D activity overseas. The gap is narrowing and if one takes into consideration the greater communication costs involved, overseas R&D activity is very likely to prove more expensive for those corporations based in the more affluent, industrialized countries.

The overseas R&D activity of corporations active internationally will probably increase in general as they try to capture larger market shares by introducing products and processes more appropriate for those markets. Such expansion of overseas R&D activity may cause very real—and sometimes costly—coordination and communication problems between corporate headquarters and overseas affiliates.

Furthermore, lack of international protection for patents and trade secrets may slow down the growth of overseas R&D activity because the initiators of a technological innovation find it increasingly difficult to exploit their new products and processes so as to profit from their R&D investment to the extent felt necessary to justify that investment.

NOTES

1. Michael G. Duerr, *R&D in the Multinational Company* (New York: The Conference Board, 1970), p. 1.
2. This paragraph and the following general discussion of arguments pro and con on decentralized R&D have been extracted from a lecture in New York city made by J. Kermit Campbell, Director, European R&D, Dow Corning International Ltd., November 1976.
3. *Business Latin America*, February 25, 1971, p. 58.
4. Stephen H. Hellinger and Douglas A. Hellinger, *Unemployment and the Multinationals* (Port Washington, N.Y.: Kennikat Press, 1976), p. 86.
5. Ivan Ivanov, "Transfer of Technology: Your Own R&D Is the Key," *Development Forum*, vol. 5, no. 3, April 1977, p. 3. Ivanov was Chief, Economics Division, Institute of U.S. and Canadian Studies, Moscow and an UNCTAD (United Nations Committee on Trade and Development) consultant at the time.
6. Government of Canada, Department of Industry, *Annual Review, 1967* (Ottawa, 1968).
7. Thomas J. Allen, James M. Utterback, Marvin A. Sirbu, Nicholas A. Ashford, and J. Herbert Hollomon, "Government Influence on the Process of Innovation in Europe and Japan," *Research Policy*, April 1978, p. 132.
8. *Ibid.*, p. 130.
9. Summarized from Allen, et al., *op. cit.*, pp. 130-145.
10. Robert M. Pierson, "R&D by Multinationals for Overseas Markets," *Research Management*, July 1978, p. 19. See also Sanjaya Lall, *Multinationals, Technology and Exports* (New York: St. Martin's Press, 1985) Chapter 7 for a discussion of technology development in LDCs and the activity of foreign-based firms.
11. This percentage is my own guess, but see Lall, *op. cit.*, pp. 38-39.
12. Jack N. Behrman and William A. Fischer, "Transnational Corporations: Market Orientations and R&D Abroad," *Columbia Journal of World Business*, vol. xv, no. 3, Fall 1980, p. 60.
13. *Japan Economic Journal*, September 13, 1986, pp. 11-13.
14. Robert Ronstadt, "International R&D: The Establishment and Evolution of Research and Development Abroad by Seven U.S. Multinationals," *Journal of International Business Studies*, vol. 9, no. 3, 1978, p. 10.

15. Hing Kwok L. Woo, "Overseas R&D of Multinational Corporations," unpublished paper prepared for the author, University of Hawaii, August 1983.
16. Ronstadt, *op. cit.*, p. 122.
17. Woo, *op. cit.*, p. 4.
18. *Ibid.*, pp. 5-6.
19. Ronstadt, *op. cit.*, p. 38.
20. *Ibid.*, p. 42.
21. *Ibid.*, p. 48.
22. *Ibid.*, p. 50.
23. *Ibid.*, p. 15.
24. Van R. Rumber, "Multinational R&D in Practice: Chemago Corporation," *Research Management*, January 1971, p. 49.
25. Woo, *op. cit.*
26. Edwin Monsfield, David Teece, and Anthony Romeo, "Overseas Research and Development by U.S.-Based Firms," *Economics*, May 1979, p. 113.
27. *Ibid.*, p. 114.
28. Jack N. Behrman and William A. Fischer, *Science and Technology for Development* (Cambridge: Oelgeschlager, Gunn & Hain Publishers, 1980), p. 2.
29. Ronstadt, *op. cit.*, p. 20.
30. Behrman and Fischer, *op. cit.*, p. 40.
31. *Ibid.*, p. 45.
32. Behrman and Fischer, "Transnational Corporations . . . ," *op. cit.*, p. 60.
33. Susumu Watanabe, "Multinational Enterprises, Employment, and Technology Adaptations," *International Labor Review*, November 1981, p. 63.
34. Behrman and Fischer, "Transnational Corporations . . . ," *op. cit.*, p. 27.
35. *Business International*, November 16, 1979, p. 365.

CHAPTER 9

Protecting Internationally Transferred Technology

For technology that is neither under patent, contained within a copyright or trademark license, nor protected by secrecy, there is no real protection other than the leverage provided by an ongoing stream of innovation. Indeed, if a firm is involved heavily in R&D and, as a result, is churning out a flow of commercially valuable innovation of product and/or process, the mere promise of the availability of the next-generation technology may be quite adequate to protect that which is currently being transferred. If appropriate protection is not maintained by the recipients, it may be understood that the next generation of the relevant technology will not be made available. But, barring that enviable position, the firm really has no effective protection for its technology unless it is patented, covered in trademark or copyright license, or maintained as a secret. All may give rise to legally enforceable monopoly of use such that the owner can impose restrictions on the use of the protected property when transferred to another party. If not, any restrictions imposed by the technology transferor on the recipient in respect to the use of the transferred technology are probably not legally enforceable. But even so, the protection may be very sparse in the case of international transfer for a variety of reasons. In this chapter, we consider each of the four forms of protection, then various international arrangements and the special problems arising in the non-market countries and

LDCs. But first, a word about the distinction among the four types of protection which, in some instances, is not all that clear.

The idea of a *patent* is to protect an inventor so that he or she can profit appopriately from the invention, thereby presumably providing an incentive for innovation. A legal monopoly is thus awarded. A patent covers an idea which is novel, significantly different from presently known technology, is non-obvious, and has commercial value. The legal validity of a patent may rest on the right of prior discovery or registration, or both. A *trademark* is essentially a certification of origin, thereby telling the customer something about product quality. The idea of a trademark had its genesis in the craftsman's distinctive mark which identified the item with a particular individual, whose reputation could add value to the product. The validity of a trademark (or trade name) arises either from prior use in trade and/or registration. The *copyright* provides protection for a particular form of expression, whether it be in the written word, picture, sculpture, phonograph record (films, program rebroadcast, videotape, product design, voice tape, or perhaps even computer software).[1] A copyright's legal validity arises either from prior authorship and/or registration. The *trade secret* relates to commercially valuable technology, information, or way of doing something, over which secrecy is maintained, either by an individual or by the organization laying claim to the substance of the secrecy. The valuable secret is often considered property under law, and its unlawful use gives rise, if stolen, both to civil suit for damages and criminal prosecution.

Obviously, there are problems in each case with respect to definition and legal basis. For example, common law countries tend to give more weight to right of prior discovery, prior use in trade, and prior authorship than do countries with a civil law tradition. This latter group tends to give greater weight to the fact of registration and the maintenance of such registration.

PATENTS

Possibly the most important device for protecting internationally transferred technology is the patent. The initial requirement is to have a valid patent in the country from which the technology is being transferred; secondarily, to assure enforcement of that patent in the recipient country.

Virtually all countries award patents only after the idea or product for which protection is sought has been run through several tests. The first is the test of novelty; that is, the idea must not have been published anyplace, domestically or abroad. The United States accepts a patent application at any time up to one year after publication, although many countries do not. That means that if a patent has actually been issued, another country may refuse protection on the grounds of prior publication. The point is that the issuance of a patent is held to be publication in that the idea is then on public view, though no one other than the inventor may be able to use it for commercial gain. The test of novelty also means that no one else has come up with the same or similar idea. (How similar?—a good question.) The requirement means that a search must be made by the relevant patent office, which in some countries is lengthy and thorough, taking several years, and in others much briefer. The products of the former are known as "strong patents," which are difficult to challenge effectively, and the latter as "weak patents," the challenge of which may be somewhat easier. Indeed, one may not know if the patent protection awarded will stick or not until after a certain period of time has passed without effective challenge to the patent. Patents issued by the United States and West Germany are generally recognized as strong; those issued by countries such as France, Italy, Luxemburg, and Belgium, as relatively weak.

The second test normally applied to a patent application is that of non-obviousness. The question is, of course, non-obvious to whom? The ordinary person in the street? Or to a reasonably well-informed person involved with the relevant technology? The third test is that of demonstrable *commercial* value. A judgmental factor, therefore, may enter the picture.

In many countries, patents are thrown open to public inspection and possible challenge before or immediately after a patent is issued. In the United States, applicants may grant permission for a valid application to be published before the patent grant, which in many cases may be worthwhile, in order to build a better case for prior discovery. Or, they may request that confidentiality be maintained until the patent is granted. The term of patent protection varies from five to twenty years. In the United States, the period of protection runs seventeen years from patent issuance. "Patent-pending" protection may run as long as five or more years, which is the period between acceptance by the patent office of a valid application (one which seems to meet all the necessary tests, for which all required documentation has been supplied, and is accepted for processing by the patent authority) and the award of a patent. The United States prohibits the foreign filing for an invention made in the United States, without first securing a foreign filing license from the U.S. Patent and Trademark Office, unless six months have elapsed since filing a U.S. application. The idea is to protect against transfers of technology possibly damaging to national security. Transgression of this requirement can lead to either loss of the U.S. patent right or criminal prosecution or both.

Some countries refuse to patent products and/or processes relating to certain types of technology. Under Mexican law, for example, patents will not be issued relative to pharmaceutical chemicals, food processing, agricultural fertilizers, pesticides, herbicides, pollution control, or nuclear energy. Sometimes, the product may be patented, but not the processes—or the reverse. In the United States, certain inventions in the atomic energy area, antipollution devices, methods of doing business, and medicines that are more in the nature of mixtures, may not be patented. To what extent patent protection may be given for living mutants (as in plants) and computer software is not entirely clear.

A number of countries require the licensing of a patent if it is deemed that local manufacture is not sufficient to meet local demand. This system of "compulsory licensing" is quite common in British Commonwealth countries. Quite frequently, the foreign patent holder has two or three years of grace, plus two more either to work the patent locally by using it in one's own facilities or to license it to another party. In certain Latin American countries, active steps must be taken to interest second parties to take a license under a patent—sort of an affirmative action program. If not, one can lose patent protection in that country.

Quite apart from any international agreement on the subject, some countries issue what are known as "confirmation patents," which are issued if applied for within a certain period by the owner of a patent issued elsewhere. The basis for issuance in such case is a valid patent in another country. One also runs into what are called "patents of introduction," which are short-term patents for something already patented elsewhere. These short-term patents are given to the foreign patent owner or, after a certain time, to a local national. The idea here is to encourage the working of the patent locally via local production. For many LDCs, from which very little technological innovation of a patentable nature originates (though this generalization may not be as valid today as a few years ago), the primary concern is to force the use of patented technology locally and to prevent its use simply to establish a monopoly within the LDC market. If full protection is given by the LDC under a valid patent, and there were no pressure to use it locally, it may be employed by the foreign owner simply to block anyone else from producing the product or using the process locally, thereby restricting local economic activity. On the other hand, if the LDC were not to give *any* protection, such fact might well discourage technology owners from participating at all in the local economy or transferring the technology to second parties. Therein lies a very real dilemma. A halfway measure is to bar the patenting of certain technologies (as in a previously cited Mexican case) or to enforce compulsory licensing.

TRADEMARKS

A trademark may be a very important asset if it is well established in the trade. Its use may facilitate market entry of a product or service by reason of greater consumer acceptance. Trademark is a generic term covering four different categories: (1) the "true trademark," which identifies a product with a specific firm; (2) service marks or names, which identify a service (a "soft" product as opposed to the "hard," an example being "Arthur D. Little."); (3) certification marks, which indicate a certain quality (examples: Underwriters' Laboratory stamp of approval or the *Good Housekeeping* seal of approval); and (4) collective marks (such as membership in an association of some sort, as a graduate of the Massachusetts Institute of Technology or the use of its logo). Protection of a trademark may run indefinitely as long as certain conditions are fulfilled.

Common law countries, as the United States and members of the British Commonwealth, frequently require continuous prior use in trade by the applicant as a condition for issuing a valid trademark registration. The civil law countries, such as many Western Europe countries, base protection directly on the act of registration. The result is that a firm may lose its trademark if it fails to register it and to maintain its validity, possibly by payment of an annual fee. Demonstrable prior use may not be adequate for restoration of ownership. The U.S. government, because of the constitutional restraint on federal powers in the field of commerce, can issue a trademark only if it has been used in interstate or foreign trade. Continuity of prior use, even if not registered, may provide adequate protection—the so-called "common law trademarks." Trademarks registered elsewhere, even though not used in U.S. trade, are recognized by the United States.

Clearly, not every proposed mark or name can be protected. The use of generic terms is out. Indeed, a mark or name may be lost if the word becomes part of the common vocabulary (examples: linoleum, cola, aspirin, cellophane, thermos, escalator, yoyo, nylon, zipper, kerosene, brassiere, trampoline). In the United States, the Federal Trade Commission may determine that a trademark has become a generally used description for a certain category of goods or services. If not, it can order the trademark null and void and, hence, usable by anyone. Also, marks should be such as not to be confused with those already existing (example: the AMF triangle in Australia where AMF is closely associated with the Australian Military Forces), or those which are offensive to good taste and public morality.

The British "Trade Marks Act of 1938," which is followed by many Commonwealth countries, created a statutory procedure for recording "registered users." This phrase refers to someone other than the owner of the trademark who has been recorded under the act. Registration follows examination by the Trade Mark Registrar to ascertain whether adequate quality control is or will be exercised by the owner over the manufacture of the relevant goods by the licensee, that is, by the registered user. Quality control may be exercised either on the basis of license agreement or on the basis of corporate control, as in the case of a parent-subsidiary relationship. The trademark protection remains valid only so long as the owner can demonstrate that adequate quality control is, in fact, maintained over the manufacture of the product by the registered user (which can be either a licensee or the local subsidiary of a foreign firm).

COPYRIGHTS

As indicated, the copyright gives protection for literary and artistic expression. It does not protect ideas, only a particular expression. Nonetheless a protectable work must be original. The U.S. Copyright Act of 1976 did away with common law protection and brought all copyrights under the umbrella of federal statutory law. The rights stemming from copyright ownership now arise from the "moment of creation." All works which qualify for copyright are protected even if not published. Some indication on the work of the fact of copyright protection, the year of first publication, and the author's or creator's name are generally required, but formal registration is optional. Registration is, however, desirable so that the copyright owner can establish a legal basis for bringing infringement action and enjoying other statutory benefits. Protection is based on where a work was first published or created, whether in the United States or in another nation, if the latter was a member of the Universal Copyright Convention at the time. Sound recordings, audio-visual works, computer programs, and data bases are specifically listed as copyrightable. The term of a U.S. copyright is now comparable to that of the Berne Convention signatory nations (to be discussed later in this chapter). In most cases, U.S. protection runs from the moment of creation to fifty years after the death of the author or creator or, in the case of corporate ownership ("works made for hire"), for seventy-five years from the date of creation. The "manufacturing clause" of the 1909 Copyright Act was extended for some years after the January 1, 1978, enactment of the 1976 act, but expired in 1986. There is now no restriction on copyright protection available under U.S. law to works by U.S. citizens or residents manufactured abroad and imported into the United States. Many countries, the United States included, prohibit the import of works bearing false notice of copyright or of any "pirated" work (one published or copied without authorization of the copyright owner).

TRADE SECRETS

The fourth category of proprietary rights rests on trade secrets, which, in common law countries, are generally treated as though patented. That is, such a secret may be the subject of a license in which restrictive covenants may be legally embedded. But not all countries recognize or enforce rights based on trade secrets, a case in point being Brazil, although the divulging of a commercially valuable secret in Brazil by an individual to an unauthorized person is a punishable offense.

Nonetheless, the laws of many countries do give some degree of protection to trade secrets similar to that accorded patents, but only so long as the secrecy is maintained. Companies may not choose to patent for a number of reasons, such as the time and cost involved in patenting, the difficulty of monitoring patent infringement and enforcing patent rights against alleged infringers, and the uncertainty of patent validity. It has been reported that well over half of all patents issued in the United States are ultimately declared invalid. Coca-Cola is an example of a company which has gone the trade secret route in protection of its "ingredient X." Over twenty U.S. states have laws that impose criminal sanctions against the unauthorized taking of valuable secret information. However, in a 1973 decision, a U.S. court held that valuable secrets that have not been patented, but which might have been appropriate for patent protection and have been in commercial use over a year, cannot be protected legally even against wrongful appropriation through breach of obligation of confidence. (Unlike patents, there is no statutory provision at the federal level in the United States for the protection of trade secrets.) Under this ruling, foreign companies are likely to think twice before licensing unpatented trade secrets in the United States, since their licensees may not be accountable for a breach of confidentiality, and a licensee's employees might be hired away by a competitor. Likewise, a U.S. company is unlikely to enter into many licenses based on trade secrets with foreign firms if the latter know that by hiring away key employees from the U.S. firm, they can caputre the secrets without committing themselves to paying royalties. There is also the possibility that U.S. firms may be tempted to transfer R&D activity to foreign subsidiaries in countries with stronger laws protecting trade secrets, such as Canada, Japan, and much of Europe.

In protecting trade secrets, companies rely upon employee secrecy agreements, such as non-disclosure and non-use clauses in employment contracts. Whether a company requires all employees to sign contracts containing such clauses, or only designated categories of employees, is a matter of company policy. In France, for example, collective labor agreements within a particular industrial sector may contain provisions prohibiting employees from committing trade secret abuses.

INTERNATIONAL PROTECTION SYSTEMS

Thus far, we have discussed only *national* protection afforded patents, trademarks, copyrights, and trade secrets. Noting the frequent mention of differences in such national laws and practices, one is moved to consider the advantages of international systems of protection.

The first such system considered here is the Paris Convention for the International Protection of Property Rights (1883), and its subsequent amendments. Basically, it is a convention to which some ninety or so countries adhere, which provides for (1) national treatment of foreign-owned property and (2) recognition of a grace period for the filing of local patent and trademark applications. In the case of patents, if a patent application is filed in one member country, the applicant may file in other member states within one year of the original filing, and the date of priority over others runs from the date of the first filing. Of course, one may still file after one year, but so may others. And, if the patent application is on public view in the patent office of the country receiving that application, which it probably is, there is a very real probability that another firm may apply. But in any event, one cannot normally file after publication anywhere, which means that the actual issuance of a patent in one country would be a bar to filing in any other. In certain countries, however, only publication in *that* country constitutes a bar to filing. One has to be wary. Note that any patents issued under the Paris Convention are national patents; one merely has more time before taking up the option of whether to apply for such protection.

In 1978, the European Patent Convention became operative, the membership of which consists of eighteen states (the twelve members of the European Community plus Liechtenstein, Norway, Austria, Sweden, Switzerland, and Monaco). It is essentially an application system, in that one may make single application to the European Patent office in Munich or directly to its search organization, the International Patent Institute in The Hague. Publication and issuance of a patent gives *national* treatment in the countries within which protection is desired. The system has the advantages of requiring but a single application and the economy of a single processing and search organization. But, in the final analysis, the applicant holds the equivalent of *national* patents in the member countries designated by the applicant, which means that the protection gained may differ substantially from country to country.

Simultaneously, an effort began to create a single European Community patent, toward which end a Community Patent Convention was drafted. Although still not operational as of early 1988, the prospects were that it soon would be. The basic notion embedded in the convention, which would apply to the twelve EC member countries, is that a single application might be made, either to the European Patent Office or to its search organization, the International Patent Institute. (Eventually, this latter organization is to be responsible for all searches for both the European Patent Convention and the Community Patent Convention.) Under the Community Patent system, a single patent would be issued covering all of the member states. The protection afforded by such a patent is to be at least equivalent to those issued under the national laws of the member states. Holding up the final ratification of the convention is the need for the member states to harmonize their patent laws sufficiently so as to remove any direct conflict among them.

The underlying purpose of the Community Patent Convention is to abolish territorial limits for the marketing of patent-protected goods. Hence, the division of the Community into national markets based on patent rights would no longer be possible. Apart from the problem of effecting an adequate harmonization of national laws, there is also the problem of circumventing the sanctity of valid patents, as against the importation of infringing goods, without prejudicing the right of a patent holder to protect industrial property as implied in the Rome Treaty (that which established the European Community in 1958). In any event, a rather long—five to ten years—transition period is anticipated.

One other international effort needs mention, the Patent Cooperation Treaty, which was ratified by the United States in 1975 and became operative in 1978 with the membership of twenty-eight countries (since expanded). This treaty does not create an international patent system, only an international patent *application* system. It provides that one patent application will serve as the national application in all signatory nations in which the applicant desires protection. Any national office may receive an application and process it, and make the search with respect to novelty. The grant of a patent is tantamount to giving the applicant the protection offered by a national patent in all of the member countries designated by the applicant. The system is administered by the World Intellectual Property Organization in Geneva. It is anticipated that eventually an individual or firm will be able to apply for either a Community patent or a European patent via the same system.

It should be noted that in ratifying the Patent Cooperation Treaty, the United States made two important reservations. First, the United States retained the right to reject the preliminary examination rights made by other countries and to undertake its own search. Second, the United States did not accept the treaty's provision that information in applications should be published within eighteen months from the priority date (even though no patent had been issued by that time). The United States did not accept this provision because it conflicted with the U.S. principle that applicants have the right to keep their inventions confidential until patent protection is actually obtained.

Although national treatment of patent rights is guaranteed by the Paris Convention, such is not always the case. For example, in the early 1980s Japan's Ministry of International Trade and Industry announced that priority rights would be granted to *domestic* enterprises in respect to the licensing of patented or patent-pending ideas for joint development of integrated circuits by the Japanese government and domestic industry.[2]

The international registration of trademarks is somewhat easier than for patents. The Madrid Agreement of 1891, in which twenty-one countries participated (but not the United States,[3] the United Kingdom, Japan, and the Scandinavian countries), provides for the automatic extension of trademark rights to all members so long as the applicant has a permanent establishment or an incorporated entity within a signatory country. The mark is communicated to the World Intellectual Property Organization in Geneva. If no duplication is found, or objection registered, internationally valid protection is given. The 1883 Paris Convention, mentioned previously, gives a six-month period of priority for filing for trademark registration in all member states after the date of filing in any one. In that the United States is a member, U.S.-domiciled firms or citizens may avail themselves of this provision.

A more recent effort in the trademark area is the Trademark Registration Treaty, to which the United States and some fifty other countries propose to adhere. This agreement would make possible a single application to the World Intellectual Property Organization in Geneva. No prior use requirement is suggested and, indeed, use would not be required until three years after international registration. National treatment would be given the trademark holder provided no member nation refused registration within fifteen months of application. For reasons already given, it would seem that there might well be constitutional basis for doubting the validity of the treaty in the U.S. case.

International protection of copyrights rests on the Berne Convention of 1886 for the protection of literary and artistic works. Membership includes some sixty countries, but not the United States. Provisions of the convention automatically extend the protection accorded local citizens or locally domiciled companies to the citizens and companies of other member countries. The period of protection is for the author's or artist's life plus fifty years. The Universal Copyright Convention (UCC) of 1955 (some fifty countries, including the United States) gives national copyright protection, which of course varies country to country. The duration of protection in some countries, in fact, is quite short (ten years in Turkey for translations). The UCC is administered by UNESCO.

There would seem to be no effective international system for the protection of trade secrets.

Efforts to protect internationally transferred technology and associated proprietary rights implicit in patents, trademarks, copyrights, and trade secrets have given rise to several important problems. These problems include those associated with trade in counterfeit goods, with the centrally planned or non-market economies, with the less developed countries, and finally, with the management of the transfer process itself.

TRADE IN COUNTERFEIT OR PIRATED GOODS

Current negotiations under the General Agreement on Tariffs and Trade, the so-called "Uruguay Round," are attempting to deal with the apparently growing trade in goods either incorporating proprietary technology without benefit of patent license, carrying unauthorized marks, or reproduced in the absence of copyright license. Countries from which complaints are arising make them on several grounds. First, such products are often of lower quality than the genuine goods and may even present a danger to health and safety for which the unlucky consumer has no legal remedy. Such counterfeiting or pirating can lead to direct loss of sales and profits by producers whose investment has generated the technology or market credibility from which others are profiting illegally. In general, the expense of identifying infringers—which is largely on the shoulders of the allegedly injured parties—is seen as unduly burdensome and the penalties, even when a case is made, much too light to be effective deterrents. The upshot is, the complaints run, that international trade is seriously distorted. For example, it is claimed that in some regions pirated sound and video recordings may represent 80 to 90 percent of the recording market.[4]

The United States has suggested various measures to protect against the trade in counterfeited goods. First, a competitive U.S.-based industry would no longer be a precondition for securing relief for patent, trademark, and copyright infringement. Second, the import of products made by using a U.S. process patent abroad without permission of the owner would be blocked. Third, the patent term for certain agricultural and chemical products would be extended so as to compensate for the time lost in obtaining U.S. regulatory approval for production. Fourth, the rights of domestic patent owners would be enlarged so as to give them the capacity to "develop innovative arrangements with potential licensees by making it clear that certain licensing practices cannot render a patent unenforceable on the ground that it has been 'misused' unless the practice also violates the antitrust laws."[5] Fifth, it would be made clear that licensing arrangements challenged under antitrust law must be judged by the actual competitive effects. That is, so-called *per se* violations would not apply. Finally, an effort would be made to clarify the rights of those licensing patents and licensees to royalty payments in disputes over a patent's validity. That is, the obligation to pay royalties would continue during the time the validity of a patent was being challenged. It has been further suggested that countries not protecting U.S. copyrights, trademarks, and patents be denied tariff concessions by the United States.

It is claimed that American industry loses more than $1.3 billion annually from the failure of ten countries to provide adequate and effective protection of U.S.-copyrighted works. The ten: Singapore, Taiwan, Indonesia, Korea, the Philippines, Malaysia, Thailand, Brazil, Egypt, and Nigeria. Additionally, it is alleged that the American agrichemical industry's annual losses from ineffective protection come to $200 million. All told, it is reported, something between $6 billion and $8 billion of domestic and foreign sales by U.S. firms was lost in 1982 because of foreign counterfeiting.[6] There seems no reason to believe that the loss has declined since then.

TECHNOLOGY PROTECTION IN CENTRALLY PLANNED ECONOMIES

The Soviet Union, the People's Republic of China, Cuba, and the countries of Eastern Europe face a special problem in that so much of the economic activity in these countries is undertaken directly by the state. Few inventors, if any, would be self-employed or in the employ of private enterprise. Therefore, ownership of virtually all inventions resides in the government. Characteristic of these countries is the issuance of an "inventor's certificate." Although the government may remain the owner, it undertakes an obligation to use the invention and to share the benefits in some stipulated percentage with the inventor (an individual or an enterprise) or pay up to some specified maximum amount.

Patent protection, although generally available, may not be important in these countries for the foreigner so long as there is no private business. Rather, a foreign firm may use a contractually based commitment to secrecy and restricted dissemination.

Typically, the foreign patent holder would be remunerated via fixed-sum installment payments rather than by a percentage-of-sales royalty. There are not only the twin problems of inspection and auditing of the patent user's books, which is often not permitted, but also a percentage of sales may well mean little or nothing where price is administered and not a function of the relatively free interplay between supply and demand. Until quite recently, protection by contract was the only method really open to foreigners in the People's Republic of China in that Chinese patents could only be held by Chinese citizens or resident aliens. The situation changed as of March 1984 so that currently foreign firms and nationals can be awarded Chinese patents.[7]

The Soviet Union, in addition to investors' certificates, also issues certificates of innovation, under which the inventor is given a percent of the savings realized by reason of the innovation, up to a specified sum. A "discovery diploma" may also be awarded, which covers any "basic change in the level of knowledge." The innovator may be rewarded a fixed amount.

Under a 1967 amendment to the 1883 Paris Convention, known as the Stockholm Convention, the inventor's certificate issued by a centrally planned economy was accorded the same status as a patent for the purpose of establishing the priority of filing dates within member countries. The United States recognized this amendment in 1973.

It should also be noted that the Soviet Union joined the Universal Copyright Convention in 1974. Some observers felt that its adherence was motivated by the desire on the part of the Soviet government to block the publication abroad of works by dissident authors rather than promote an exchange. The People's Republic of China enacted a trademark law in 1982.

THE PROBLEM OF TECHNOLOGY PROTECTION IN LESS DEVELOPED COUNTRIES

Understandably, many LDCs have had serious second thoughts about the value of a full-fledged patent system to them, given the fact that the overwhelming percentage of all patents issued are owned by individuals and corporations domiciled in the industrialized countries. Even a country such as Canada, where 95 percent of all patents issued in the country are based on inventions by foreigners, has become suspicious that the patent system may be serving as a brake on Canadian industrial development. A number of proposals have been made in Canada, including reduction of the period of patent protection from seventeen years to nine (with an added five years for those actively working their patents in Canada), requiring greater disclosure of information by patent holders, changing from a "first to invent" system of priority rights to a "first to file" system, and allowing imports of a patentee's product made outside Canada to be marketed freely in the country, thus discouraging non-competitive pricing in Canada. One result has been introduction of compulsory licensing in certain sectors.

Many LDCs object to the traditional notion of a patent because:

1. Patents guarantee national treatment to foreign patent holders. Some have suggested that foreign-owned patents should be treated differently from locally developed patents with respect to duration of protection and compulsory licensing requirements. The rationale advanced for the difference is threefold: (1) R&D allocation in the industrial countries is not influenced by patent protection in the LDCs and, hence, does not provide incentive for innovation;[8] (2) most patents issued by the LDCs are never

worked in the country; and (3) many patents are simply used to exclude competitors from LDC markets.

2. Patents are often not used. Under present laws, even where compulsory licensing is enforced, it takes too long to compel licensing (one-year priority for local filing, plus three years during which the compulsory licensing rule is not enforced, another two to three years to establish liability for not working the patent, plus perhaps another two years before a local court finally orders compulsory licensing). By that time, at least half of the life of the patent has expired, and the technology may well be outdated.

3. A patentee has a right to exclude imports of a non-licensed manufacturer with respect to a patent protected locally, even though no local production takes place. Some have urged that such protection should be limited only in cases where there is actual production of the patented product.

4. No criteria are considered concerning how patent protection contributes to a country's economic development when a patent is issued.

5. Public welfare may be ill-served by covering such essentials as pharmaceuticals and food products by patent; many LDC governments feel such protection is intolerable.

6. A foreign parent corporation can license a patent to a local firm over which it exercises control, in which case the extraction of royalties is seen by the LDC simply as a device to drive down local profits and, hence, local tax liability.

The patent issue has entered the North-South dialogue, and the World Intellectual Property Organization has urged a number of revisions in the Paris Convention along some of the lines suggested above.[9] Some have argued for the abolition of patents altogether. But if there were no patents, inventions could be kept secret—and hence, proprietary—much longer, and thus could not be used by others as a basis for further research. As noted earlier, once a patent is issued, and often before, application or patent documents describing the technology in detail are generally available to anyone for a nominal fee.

In any event, there is an admitted asymmetry in the situation. In recognition of this asymmetry, Japan (in 1984) offered its government-owned high-tech patents free of charge to interested LDCs.

Trademarks pose a somewhat different problem. The use of an internationally known trademark or name by an LDC firm may be highly beneficial in gaining access to both the local and international markets. However, if the LDC firm is restricted so that it can only use the established foreign mark or name, the use of which it has licensed, it becomes captive to the foreign firm owning that mark or name; the LDC firm has developed no mark or name of its own. To offset this problem, the Mexican government has, for example, stipulated that any Mexican firm licensing a foreign mark or name must also use its own mark and name in an equally prominent way, whether on products for local consumption or for export. In that way, over time the local firm can wean itself from the foreign mark or name and go with its own. On the other hand, some foreign firms may not permit its local LDC licensee to use the local name or mark in external markets out of fear that product quality will not be maintained by the LDC manufacturer. The non-use of the foreign mark or name may render it virtually impossible for the LDC firm to compete on foreign markets, which, of course, may be the real purpose of the licensor.

Copyrights pose still another problem. It may be deemed in the interest of the LDC to translate and copy as many foreign books and publications, tapes, films, computer software, etc. as possible without paying any royalty. The transfer of technology in the public domain is thus facilitated, which may be seen as highly desirable. Furthermore, the typical LDC may produce little for which it desires foreign copyright protection. Hence, no reciprocity is involved. Stealing from affluent countries and corporations abroad is not seen as reprehensible. One is driven by the moral duty to capture foreign technology as cheaply as possible and thereby stimulate local development. Without excusing, one can understand.

The transfer of technology into LDCs via license based on secrecy is likewise a problem. Is it really in the interest of the receiving country to make that secrecy enforceable? One of an LDC's purposes in securing foreign technology is its rapid dissemination throughout its society, thereby stimulating more rapid modernization of productive activity. Generally, it is probably a safe observation that the laws in many LDCs are not conducive to the effective enforcement of secrecy clauses in employment contracts or of non-disclosure clauses in technology transfer agreements. Very frequently, even where a contract carries such a non-disclosure covenant, it becomes explicitly non-operative if the government directs otherwise, which is not infrequently the case.

MANAGEMENT PROBLEMS

Major management problems arising out of international efforts to protect rights based on patents, copyrights, trademarks, and commercial secrets transferred internationally are manifold. Some of the more important arise because of:

1. The need for an early evaluation of the commercial value of a patent in that the first party to apply for a patent in one Paris Convention country has only one year of grace before applying for protection in other member states (That is, it has priority against others during that year. It can, of course, apply later if no one else has applied in the meantime, so long as publication has not occurred. As pointed out earlier, the actual issuance of a patent anywhere is equivalent to publication and hence makes patent coverage elsewhere impossible.)

2. The need to set up a system to assure continuation of patent, trademark, or copyright validity (Payment of an annual fee may be required, and, in the case of a patent or trademark, local use may be required; for a trademark, periodic application for renewal may be required.)

3. The need to set up a system to monitor important markets for infringement and to prosecute infringers

4. The need to decide whether to develop a universal trademark or a series of national or regional marks (If the former, the problem of avoiding legal restrictions and unpleasant connotations in every country is a formidable one requiring substantial research.)

5. The need to decide who should own the proprietary right—the parent company or subsidiary (If the parent is to remain the owner, the subsidiary must be licensed to use the right. But such licensing of subsidiaries is not legal under all systems. If the sub-

sidiary is 100 percent–owned, the only difference is that between a dividend (posttax earnings) and a royalty (a pretax cost). But if it is a joint venture, the local partner may not take kindly to payment of a royalty (which is a cost) to the foreign partner, particularly if the level of the royalty is determined by that firm. On the other hand, if a proprietary right is sold to or capitalized by a subsidiary abroad, the parent must assure itself that the right is not lost to it if ownership of the subsidiary changes hands.)

6. The need to have continuing access to legal counsel expert in international trademarks, copyright, patent, and trade secret law

7. In that prior use is not a requirement in many countries for registering a trademark, and in that the United States is not a member of the Madrid Agreement, the possible need for a subsidiary in one of the countries signatory to the Madrid Agreement through which to register the trademark (But in such case, the subsidiary must own the trademark—which introduces the problems suggested in 5 above.)

8. The need to compare the desirability of relying on patent protection (which constitutes publication) or on maintaining secrecy (hence, denying any knowledge to competitors)

Regardless of the problems involved in protecting intangible proprietary rights abroad, many thousands of firms have found the transfer of such rights to be a very profitable business. In the final analysis, the real protection of transferred technology lies in an untarnished corporate name, a high level of credibility in the eyes of consumers in all markets, superior technology supported by a record of ongoing innovation, a capability of effecting international technology transfer at relatively low cost, a design-engineering sensitivity to the requirements peculiar to different markets, and sufficient organizational flexibility to select the most effective transfer vehicle and mount appropriate controls.

NOTES

1. Some countries, including the United States, offer protection via "design patents," which provide a shorter term of protection than the normal pa-

tent, say five years versus fifteen to twenty years. Also, some countries issue "utility patents" for minor inventions of commercial value and give abbreviated protection.
2. *Japan Economic Review*, June 16, 1981, p. 17.
3. The United States was prevented from joining the Madrid Agreement because it could not give automatic coverage for registered trademarks in that U.S. protection was contingent upon establishing *prior use* either in interstate or international trade.
4. The *GATT Newsletter*, July/August 1987, p. 4.
5. Eileen Hill, "Protecting U.S. Intellectual Property Rights," *Business America*, April 14, 1986, p. 4.
6. *New York Times*, April 7, 1986, p. D-1.
7. The People's Republic of China adhered to the Paris Convention in 1985.
8. Very clearly not the case. Many firms undertake R&D and introduce technological innovation on the assumption that the new process or product will be introduced simultaneously in multiple markets. Otherwise, the investment in the R&D could not be justified financially.
9. Edith Penrose, "International Patenting, and the Less Developed Countries," *Economic Journal*, September 1973, p. 768 ff. Another landmark contribution to the discussion of the LDC issue is found in Constantin V. Vaitsus, "The Revision of the International Patent System: Legal Considerations for a Third World Position," *World Development*, 1976, vol. 4, no. 5, p. 85 ff.

CHAPTER 10

International Competitive Bidding

The subject here has to do with the procurement process by which governments, and sometimes private firms, acquire technology and related services and hardware from foreign suppliers, a subject but rarely dealt with in the literature of international technology transfer.*

The method of procurement favored internationally is international competitive bidding (ICB), also known as "public bidding," "open tendering," or *"apel d'offres."* Major multilateral development finance institutions, such as the International Bank for Reconstruction and Development (via the International Finance Corporation and International Development Association), the Asian Development Bank, the Inter-American Development Bank and the African Development Bank all prescribe open international competitive bidding procedures for most of the procurement for which their funds are used.[1]

So likewise do many of the national development finance institutions, some of the principal ones being: Caisse Centrale de Coopération Economique (French), Commonwealth Development Fund

---

*The basic text of this chapter was prepared by William Tan Beng Chuan and is used here with his permission. It is extracted from his advanced study project submitted in partial fulfillment of the requirements of the degree of Master of Business Administration. National University of Singapore, School of Post-Graduate Management Studies, March 1984 (unpublished).

(UK), Industrialization Fund for Developing Countries (Danish), Netherlands Finance Company for Developing Countries (Dutch), Overseas Economic Cooperation Fund (Japanese), Societé Belge d'Investissement International (Belgian), Swedfund (Swedish), and the U.S. Overseas Development Corporation. However, several alternatives to competitive bidding do exist.

A number of arguments are advanced for sourcing from abroad, among these are the following considerations:

1. The non-availability of specific goods and services in the local market, possibly because of small market size
2. The unsatisfactory quality of locally available goods
3. Unduly high prices on the local market
4. The unsatisfactory nature of post-sales services available on the local market
5. Unsatisfactory conditions of sale imposed by locally based suppliers
6. Customer preference
7. Replacement of goods of foreign origin
8. Reciprocity of trade, possibly by reason of balance-of-payments pressures or obligations under bilateral trade treaties
9. Strategic reasons, such as assuring greater security of supply by developing a source in a different country [2]

Technically, procurement is distinguished from the simple purchase of goods in that the former includes the acquisition of both goods *and services*, for example, the hiring of contractors or consultants to work on a project. Hence, procurement always involves technology transfer in its true sense. The functional scope of procurement covers:

1. Specification of the kind and quantity of goods or services to be acquired
2. Investigation of the market for supply, and contacts with potential suppliers
3. Placing the order or contract, including negotiation of terms
4. Supervising delivery and performance
5. Taking necessary action in the event of inadequate performance
6. Payment
7. Dispute resolution [3]

ALTERNATIVE METHODS

Alternative methods of procurement include selective bidding, single-source procurement, negotiated procurement, and competitive bidding. In selective bidding, also known as "limited tendering," bids are invited directly from a preselected list of suppliers. Single-source procurement, as the name implies, means that the buyer approaches a single supplier. Negotiated procurement refers to a system in which the purchaser first chooses a supplier or contractor satisfying criteria other than price and then attempts to negotiate a mutually satisfactory contract, if possible. If that proves impossible, a second supplier is approached. All of these methods have obvious disadvantages and are rarely approved by governments or by international financing institutions, although frequently used by private buyers.

Reasons for the widespread use of international competitive bidding are essentially three; (1) a public agency is bound to provide an equal opportunity to all potential bidders. (It is frequently under treaty obligation to show non-discrimination vis-a-vis countries and companies); (2) competition results in the most economic use of public funds; (3) the open competitive bidding procedure serves as a safeguard against waste, corruption, and favoritism.

International competitive bidding represents an effort to reach all qualified bidders through international notification, and thereby provide them with equal opportunity. Standards and specifications are selected so as to not discriminate against or favor any company. Precise criteria for selction of the successful bidder are specified insofar as possible. Such a process is specifically recommended by many international financing institutions and possibly even required in order to gain access to their financial resources.

THE BIDDING PROCESS

In the procurement of goods or public works through ICB, the process can be broken down into the following steps:

Step 1. *Preparation of Bid Documents.* Bid documents include an invitation to bid, instructions to bidders, condition of contract, bill of quantities, technical specifications, and a bid form. In ICB, the bid documents must be ready for distribution to prospective bidders as soon as an advertisement giving notice of the project has been published.

Step 2. *Advertising.* Critical factors here are the choice of media, content of the advertisement, and the time allowed for the submission of bids.

Step 3. *Prequalification.* It may sometimes be advantageous to prequalify firms from among those who respond to international notification, particularly for large projects, in order to determine which have the technical and financial qualifications to undertake the project. The actual bidding is then confined to these prequalified firms, that is, "shortlisted" firms.

Step 4. *Bid Invitation.* Reference here is to the formal step by which the buyer or owner invites potential bidders to submit bids. Potential bidders receive the invitation to bid and bid documents, including the instructions, conditions of contract, and technical specifications. The potential bidder may be required to buy these documents so that the expense of providing them to firms not seriously interested may be eliminated.

Step 5. *Bid Opening, Evaluation, Selection.* The evaluation period, which begins with bid opening, is an important phase of the procurement cycle. Procedures must be carefully established to ensure the fairness and integrity of the process so only the best prices may be obtained. Bid opening should be public and not subject to subsequent change by either buyer or supplier. Evaluation and selection of the successful bidder should follow criteria specified in the bid invitation, so the process will be perceived as fair by the participants (an essential feature for success of ICB).

Step 6. *Postqualification.* A contract should be awarded only to a bidder who is technically, managerially, and financially qualified to perform it. If prequalification procedures are not used, this step is accomplished by "postqualification" of the lowest evaluated bidder—and if he is not qualified, of the next lowest bidder. Information necessary for the evaluation of such qualification should be requested in the bid documents. (In some instances, governments may be required to accept the lowest bid.)

Step 7. *Contract Award.* Notification of award is followed by execution of a written contract by the two parties. The bid documents should state how the bidders will be notified of an award and at what point the owner regards the contract as legal and binding.

Step 8. *Contract Performance.* The procurement of any complex or large project involves both the procurement of works and services and the purchase of goods. In the procurement of works and services, contract performance involves construction, inspection, measurement, payments, variation orders, resolution of differences and disputes, and final payment. In the purchase of goods, contract performance includes manufacture, predelivery inspection, transportation from country of manufacture to site, unloading, customs clearance, inland transportation, erection, performance testing, payments, and warranties. The rights and obligations of both parties should be governed by the contract conditions.

Step 9. *Contract Finalization.* Finalization is the terminal phase of ICB in which the contract is reviewed to determine that all obligations have been carried out by both sides before the contract is terminated and a final financial settlement reached. In the procurement of large and more complex projects, this review would be carried out after the expiration of a guarantee period to ensure that all possible problems were solved.

When there is insufficient expertise within a government department or agency, or private buyer, to initiate and manage a project, it may become necessary to hire a consultant to assist in the procurement process. Assignments given to consultants may include the development of feasibility studies, detailed engineering and design, and supervision. Additionally, consultants are often used as advisors in the preparation of tender documents and in evaluating bids. In the past, consultants have been selected mainly on the basis of demonstrable expertise and reputation, with less consideration in respect to price. Currently, price seems to be becoming an increasingly important factor, and consultants themselves are often asked to submit proposals on a competitive basis.

NEGOTIATION

Negotiation is defined here as the process of working out a procurement and purchase to the point of reaching a mutually satisfactory agreement. But is negotiation compatible with competitive bidding? The misconception is that buyers should select the lowest bid rather than engage in negotiation. It is often pointed out that the price-based competitive bidding approach is difficult and even impossible in the purchase of many types of equipment (for example, sophisticated electronic systems that require substantial instruction and maintenance) because too many other important factors are involved in the purchase. The dangers in accepting the lowest bid are several-fold:

1. Unless the specifications are detailed and comprehensive, the winning bidder may provide a product that meets specifications but not provide the performance expected by the buyer.
2. The bidder may cut back on quality to make possible submission of a low-priced winning bid. The high-quality producer thus becomes non-competitive and exists from the market. The result is that the buyer must live with an inferior product, one which perhaps wears out or falls apart sooner than expected.
3. A bidder who succeeds in getting the bulk of a buyer's orders can possibly drive its competitors out of the field. Subsequently, it can then raise prices virtually at will.
4. Given a situation in which a few firms dominate the supplying industry, suppliers may adopt a follow-the-leader price strategy, in which case, bidding fails to produce the desired price competition.

Negotiation is desirable if the buyer is to meet the objective of getting the specified quality at at best possible price, but one is circumscribed by the requirements of the relevant law and of good business ethics. In the absence of negotiation, the buyer must assume that the bids reflect the best prices and acceptable quality, which may not always be the case. Hence, negotiating skills may be as important to the buyer as to the bidder.

THE PRICING PROBLEM

Even though the borrower and the public (or private) international lender may both wish to execute projects economically and efficiently, other considerations often conflict with these objectives. For example, the World Bank and its affiliates, many of the regional development banks, and very commonly, national development agencies, restrict procurement to member countries, or in the latter case, to national sources. Such restraint means that funds cannot be used for procurement in other countries even if they could supply the goods or services desired at a lower cost. Another consideration is the granting of domestic preferences by the procuring country. Domestic contracting and manufacturing industries are often given a margin of price preference in the evaluation of the bids, as does the United States in the public sector under terms of the "Buy American" law.

There are three general types of pricing agreements: firm fixed-price, redeterminable fixed-price, and cost-plus.

A firm fixed-price agreement is one in which a fixed consideration is established on a total contract basis. This price, once established at the signing of the agreement, remains unchanged throughout the life of the project. A price change may be justified only when the buyer wishes to change the scope of the contract or when there is mutual error. The advantages of fixed-price agreements are many. First, they are simple, both to establish and to administer. The buyer obtains maximum protection as payments are made only after delivery and proof of performance. Conditions favoring such agreements:

1. Competition is strong and prices are responsive to supply and demand.
2. Previous purchases of identical or similar equipment make for reasonable price comparison.
3. Accurate cost information and experience are available for determination of fair and reasonable prices.

On the other hand, fixed-price agreements have serious shortcomings in some situations, especially (1) where many changes are anticipated in scope, content, or mode of performance during the project (particularly relevant in the case of very complex and/or long-lived projects); and (2) where serious contingencies are present which do not permit realistic estimation or evaluation until occurrence. Examples are floods, earthquakes, high winds, draughts, power failures, strikes and civil unrest, or insistance on a use of an untested technology and its failure.

Redeterminable fixed-price agreements, or repricing agreements, occur in cases where actual cost experience is accumulated before the fixed price is finally established. This fixed price might be retroactive to the start of the contract and prospective to its completion. Although repricing agreements may be costly and time-consuming to administer, this type of pricing agreement can be applied advantageously under the following circumstances:

1. Initially, specifications are incomplete, and development and production are carried out jointly.
2. Costs (start-up costs, production costs, etc.) are difficult to estimate.
3. The complexity of the product or process would necessitate the inclusion of a large contingency fee in a firm fixed-price agreement.
4. Many changes are expected during the project.

Specific types of redeterminable fixed-price agreements are "flexible pricing" and "retroactive pricing." Suffice it to note that these types differ in terms of allowable price changes. Although a fixed fee is paid in either case, the timing and allocation of the variable add-on differs. In the final analysis, if actual costs turn out to be greater than estimated, the seller receives a lower percentage profit.

Though in general the firm fixed-price contract seems to be the best type of contract, it does not meet the problem of price escalation. Some projects take a long time to implement. The time between when a contractor calculates his bid and the commissioning of the project (and final payment made) can easily be in the order of several years. During that time, price levels may well change significantly. By requiring a firm fixed bid, it may appear that the buyer is asking the tenderer to bear all the risks of future price escalation or inflation. But, in fact, the tenderer will doubtless anticipate inflation and, looking at the cash inflows (payments) and the cash outflows (e.g., purchases) will adjust prices accordingly. The question then is: Could one have a situation in which a tenderer is really the lowest bidder but because a higher level of inflation was anticipated, arrive at a higher final price than another tenderer who actually assumed a higher price but anticipated a lower level of inflation? In any event, changes may be retroactive to the beginning of a contract (or to some defined period) and/or prospectively to the end of the contract (or to some defined period).

Cost-plus–type agreements reflect the case in which the final price is established on the basis of actual cost experience. There are two types in common use: time and material agreements and cost-plus–fixed-fee agreements. Under a time and material agreement, a predetermined time rate is agreed to by buyer and seller, covering direct labor, indirect labor, overhead and profit, also, a percentage-of-cost fee. A basis for material evaluation (list price, wholesale price or cost) is likewise established by agreement. When the work is completed, actual costs are compiled and the fee is calculated in terms of the original agreement. Care needs to be taken in such cases to avoid abuse and to assure that adequate controls are established to ensure minimum cost and to avoid wasteful methods and procedures. This type of agreement should probably be employed only after determining that no other type of agreement is adequate. Under cost-plus–fixed-fee agreements, an estimate for the "target cost" is first established, on which basis a fixed fee (profit) is established. When the work is completed, actual costs are compiled and computed, and the seller is reimbursed accordingly. In fact, it can happen that the bid is actually awarded to a contractor who substantially underestimates the inflationary factor. There is a high probability that the contractor will act in such a way as to elevate project cost, produce inferior work to cut its own costs once the error is realized, or intimate that it will experience financial failure unless the contractor's basic fee is defined as a percentage of actual project cost.

There is thus a case to be made for the buyer to absorb most, if not all, of the risks of inflation. This can be done by allowing the contractor to make price adjustments according to some price index or other objective measure, whereby the risk to the contractor is reduced substantially. The World Bank makes the following relevant observation:

> The risk does not have to be eliminated totally. In setting up contractual provisions for price adjustment, it is not necessary that absolute numerical precision should be sought in order for transfer of risk to the employer [i.e., buyer] to be achieved. The risk carried by the contractor should be reduced to the point where he believes a substantial bid risk allowance is

unnecessary. Further pursuit of accuracy is likely to generate difficulties and to absorb administrative man hours without benefit to employer [buyer] or contractor.[4]

There are two main methods of price adjustment: price adjustment on documentary proof and formula methods of price adjustment. With price adjustment on documentary proof the contractor has to show documentary proof that it has incurred increased costs in the execution of the contract because of an increase in material prices or the cost of labor. This method of price adjustment tends to lead to arguments and disputes. It also requires considerable administrative work, and contractors frequently feel that they are compensated less than the actual cost increase. Generally this price adjustment method is used only when satisfactory price indices are not available. The formula methods of price adjustment call for several indices to be incorporated into a formula for the adjustment of prices. For example, an officially published labor index and several material price indices representative of the actual materials used may be employed to build up the formula. The formula method requires minimal administrative resources and does not require records of actual cost increases. Additionally, this approach can remove the risk of inflation for the contractor with respect to all types of costs. It does depend, however, on the availability of accurate, unambiguously defined indices from mutually acceptable, published statistical sources. An example of a cost escalation formula is:

$$P_2 = P_1 \times \left(0.15 + 0.4 \frac{L_2}{L_1} + 0.45 \frac{M_2}{M_1}\right)$$

where P_1 = tendered price
P_2 = price of future orders
L = a regularly published earnings index
M = a regularly published wholesale price index for major materials used in the contract

Subscript 1 = indices on which the current tender prices are based

Subscript 2 = indices relevant to the period in which future orders are placed

The currency in which contract payments are to be made is usually agreed upon during negotiation and specified in the agreement. Because of floating exchange rates and the long duration of many projects, risk from foreign exchange fluctuations may be significant. In general, there are four ways to allocate the risks of foreign exchange fluctuations. *First*, the buyer may assume all of the risk and agree to pay a fixed amount in the currency of the seller's country. *Second*, the seller may assume all of the risk and agree to accept payment either in the currency of the buyer or at a fixed exchange rate. *Third*, the risk may be shared between the buyer and the seller in one of several ways, ranging from a percentage-sharing of the risks to a situation in which the risk for one of the participants is limited to a specified maximum. Using this latter approach, the amount of risk might then be limited to a certain exposure level for that participant. For example, only a stipulated proportion of the payment might be at a fixed amount of currency, either of the seller's or buyer's country. *Fourth*, the risk is eliminated by hedging via a forward foreign exchange contract. The cost of hedging might then be allocated between the buyer and the seller, such division to be determined during negotiation.

A contractor under the ICB process usually incurs considerable expenses before receiving the first payment for goods supplied and services rendered. The contractor may have to start up production of equipment required on a project, or subcontract it and make advance payments. The contractor's engineers may have to undertake detailed design and engineering work, put together a system and test it in the factory. The contractor may then be required to deliver the equipment to site before receiving the first payment. Hence, it is a recognized procurement practice to advance a payment to the contractor shortly after the contract is awarded. According to the World Bank supplemental guide, competition is enhanced by bringing in bidders who have a more difficult access to credit. Bid prices may then be reduced by providing the supplier with ready cash required to start work.[5] Advance payments are usually about 10 percent of the contract price, seldom exceeding 30 percent.

The World Bank recommends that certain safeguards be taken when advance payments (also, progress and time-scale payments) are made. One is a bank guarantee obtained by the buyer for the amount of the advance involved, guaranteeing return of the advance payment if the bidder is unable to fulfill the contract. A second recommended safeguard may be a provision in the contract to the effect that all materials purchased by the contractor for the project become the property of the buyer, wherever those materials may be located.

Finally, a provision may be included in the contract to the effect that the supplier shall maintain insurance in favor of the buyer against all risks of loss or damage, for the full value of all project-related materials, in process or stored.[6]

Since contracts often last up to several years, it is essential that the buyer make periodic payments to the contractor. Such payments obviously have considerable impact on the contractor's cash flow. As the contracting firm often operates on borrowed funds, the payment terms stipulated in the contract can be of signal importance to it.

Acceptance of the project occurs after tests have been conducted by both buyer and contractor and these tests show that the equipment or project performs according to specifications. A payment usually follows acceptance of the project. However, this may not be the final payment, as the buyer may still retain a certain percentage to ensure that the contractor will rectify promptly any defect which may come up during the warranty period.

A final point concerning payments should be noted. If the supplying firm is contracting with a foreign buyer with which the supplier has had no past experience, whose credit rating is uncertain, or which is located in a country in a deficit position relative to the foreign currency in which the supplier requires payment, the supplier has two possible options. *First*, it might require advance payment into an escrow account in a third country, or issuance of an irrevocable letter of credit. In either case, payment would be made to the supplier upon presentation of documents proving that the underlying contractual obligations had been completed. Such documentation would have been specified in the underlying contract. *Second*, the supplier may insure against non-payment. Such insurance may be available from an export-financing institution in the supplier's home country; for example, the Export-Import Bank, the Foreign Credit Insurance Association, or the Private Overseas Private Investment Corporation in the U.S. case. Other capital- and technology-exporting countries have similar institutions. If one is working out of a country lacking such facilities, access to another country's institutions by working through a branch office or subsidiary in that country may be important.

As unbundled technology transfer increases across borders, so likewise does the project type of transfer—that is, transfer via A&E,

construction, system installation, turnkey, and turnkey-plus types of contract. As the size and complexity of such projects grow, international competitive bidding becomes a way of life for those firms in the business of selling technology externally, to overseas entities *other than* the supplier's own branches, associated firms, and subsidiaries.

NOTES

1. Raghavan Srinivasan and David M. Sassoon, *International Contracting and Procurement for Development Projects*, Volume 1 (Washington: International Law Institute, Georgetown Law Center, 1982), Chapter 1.
2. Adapted from Srinivasan and Sassoon, *op. cit.*, Chapters 2 and 4.
3. *Ibid.*
4. World Bank, *Supplemental Procurement Guide for Bank Staff* (Washington: World Bank, 1981), pp. 7-1, 7-11.
5. World Bank, *op. cit.*, pp. 5.1-5.7.
6. *Ibid.*

CHAPTER 11

Selecting the Transfer Mode

The organizational vehicle, or mode, used for the international transfer of technology may be of singular importance to the success of the transfer. The transfer options range from *internal* transfer to the firm's own branches or subsidiaries abroad to *external* transfers via equity or contractual joint ventures and partnerships, or to completely unrelated parties via technical collaboration agreements. Using the terminology adopted in this text, all of these external transfer modes are subsumed under the phrase "joint enterprises."

The question which arises is whether a given firm proposing to transfer technology really looks at the full array of transfer options in its effort to maximize return over some period of time, or simply dismisses one or more as not permitting adequate control of the technology and its ultimate use.

The casual assumption of many corporate executives, that control of technology via ownership in the parent-subsidiary relationship is more complete and more easily enforceable than control via contract, requires examination, at least in the international case. Implicit in that assumption are two corollary assumptions, neither of which may be valid. The two:

1. that control via ownership is possible; and
2. that control via contract is unenforceable.

In the purely domestic case—that is, in a largely capitalistic society—the equation of ownership with control may be justified. Even so, however, one must be sensitive to the many areas in which political-legal, as well as social-organizational, restraints impinge on the degree of an owner's control over transferred technology. In the international case, where markets may be very disorganized, prices far removed from representing true scarcity values, and national values and priorities at wide variance, governments intervene in many ways, as we have already noted. In so doing, they impinge heavily on ownership-based control. In the polar case, private ownership of productive assets, whether the owners be domestic or foreign, is simply not tolerated. Or, only certain categories of assets may be owned privately. One trouble is that the ownership package tends to be only broadly defined in terms of what is being controlled, by whom, over what time. Contractually based control, on the other hand, is generally much more explicit on these points in that it is related to certain transfers and functions within a given period of time. Therefore, such control—particularly if the underlying contract is subject to official approval, as is frequently the case in LDCs—may be more clearly seen as justified. It is finite in all dimensions, unlike ownership-based control. Also, specific payments are more likely to be linked with specific transfers and, hence, the trade-offs may be evaluated more easily. The result is that many governments clearly prefer technology and capital transfers (and associated control) via contract to foreign direct investment and foreign ownership. Even where some equity investment is involved, as in a joint venture, the relationship is frequently in the nature of a contract in which control, payments, and duration are specified and limited. The result is that contractually based rights or control are less likely to be intervened by a government than the more broadly based and ill-defined ownership-based rights or control.

Supporting this thesis is evidence that a number of countries are demonstrating preference for "unbundled technology" imported under contract within a joint enterprise of some sort, rather than as part of the foreign direct investment package in a foreign-owned subsidiary. It has been observed that:

> As the NICs[1] progress, there would seem to be increased scope for newer, more "autonomous," forms of partnership between firms in these countries and those in the developed countries; e.g., joint ventures, licensing and technical assistance agreements. Such arrangements can provide an advantageous

route to technology transfer as well as ensuring greater domestic control over NIC industry than is possible with the establishment of wholly or majority foreign-owned enterprises.

The more dynamic NICs have resorted to these newer types of partnerships as a means of integrating foreign technologies from multiple sources so as to develop "indigenous" upmarket products. Examples abound in advanced electronics and automobiles—for instance, the use that has been made of Italian design and modified Japanese Mitsubishi engines by South Korea's Hyundai automaker, itself 15 percent owned by Mitsubishi. And Brazil's Embraer Aircraft Company has been linked up with a number of European firms under licensing agreements or joint ventures, notably for the production of light military planes.[2]

TYPES OF JOINT ENTERPRISES

We deal here with those undertakings in which enterprises from two or more countries commit capital assets (finance, access to markets, land, production equipment, technology, skills, and/or proprietary rights) to a joint enterprise, perhaps share some degree of management responsibility, and possibly participate in the commercial risk of the undertaking up to the value of their respective contributions or as defined by agreement. Typically, they are entered into for a specified number of years or until dissolved by mutual consent. The partners share in the earnings of the venture as agreed upon and also may derive benefits via other channels, such as the purchase of goods and services from the enterprise or the sale of goods and services to it under separate contract. As indicated, such ventures are becoming increasingly important as conduits for the transfer of technology. They differ substantially from the conventional technology transfer agreement in that the intent is for an ongoing joint activity in which management and risk is shared to a degree.

There is much confusion in the literature among three types of joint enterprises—the equity joint venture, the partnership, and the contractual joint venture. Legally and structurally, they are distinct, though problems of control and management may be similar. Figure 11-1 summarizes the distinctions. Bear in mind that in all cases, financial accounts separate from those of the owners or partners would normally be maintained.

The fourth form of joint enterprise is one in which the "jointness" is more narrowly proscribed, the technical collaboration agreement *standing alone*—that is, outside of any organizational structure,

Figure 11-1. Distinctions Between Equity Joint Ventures, Contractual Joint Ventures, and Partnerships.

The Equity Joint Venture	The Partnership	The Contractual Joint Venture
Based on local corporate law, corporate articles of incorporation, and by-laws	Based on partnership agreement and local law relating to partnerships	Based on contract and local commercial law relating to contracts
Separate corporate identity, own personnel (including management)	Joint management in a separate, identifiable structure, but with *no separate incorporated identity*	No separate organizational identity necessary, but may have joint management of projects and/or functions assigned to it by the partners in the underlying contract
Invests and contracts with third parties, in its own name; may sue and be sued in court of law	Invests and contracts with third parties, only as agent of the partners and as authorized by them	Investment and expenses divided and assigned according to fixed percentages as stipulated in the underlying contract; contracts with third parties as authorized
Liability limited to joint venture's assets	Liability flows through to the partners' assets	Liability of each partner limited to its investment unless otherwise specified in the underlying contract
Income taxed to the joint venture	Income taxed to the partners (is simply a conduit for revenues and costs)	Income taxed either to the joint venture or to the partners, depending on local law
Work allocated to the partners by specific contract	Work allocation agreement	Work allocation agreement
May accumulate earnings and reinvest	Earnings as assumed to have been distributed to the partners as earned	May accumulate earnings and reinvest as per underlying contract
Assets are the property of the joint venture	Assets may be declared by agreement as indivisible	Assets may be declared by agreement as indivisible
Not necessarily limited in time nor to specific objective	Not necessarily limited in time nor to specific function—only to a general purpose as defined in the partnership agreement	Limited in time and a more narrowly defined purpose (e.g., to execute a specific project)
Terminated by liquidation or bankruptcy under rules of local corporate law	Termination by withdrawal of a partner or at a predetermined time	Termination conditions defined by underlying contract (often upon completion of a project or on a specified date)

Source: Richard D. Robinson.

which is implicit in the three forms of joint enterprise charted in Figure 11-1. The two characteristics necessary are a genuine transfer of technology (as defined in the early part of this book) and a relationship which continues for a period of time. It is, of course, quite possible that one or more technical collaboration agreements may be associated with any one of the other three forms of joint enterprise, which is frequently the case. Here we are using the phrase, "technology collaboration agreement" in a generic sense. Included is any contract calling for the international transfer of technology and/or associated rights. These contracts include (1) the pure license (by which a proprietary right is bestowed on a second party, if instruction and explanation is included); (2) the sale of a capital asset (if associated with some sort of instruction and/or service agreement); (3) the technical assistance or know-how agreement; (4) contract manufacturing (if associated with technical assistance of some variety); (5) training contract; (6) consulting contract; (7) architectural and engineering contract (if the training of local personnel is involved); (8) research and development contract (if part of the R&D is performed locally); (9) construction, construction supervision, and turnkey contracts (if performed in cooperation with local personnel); (10) turnkey-plus contract; (11) production-sharing arrangement; and (12) co-production.

It is of significance in this regard to take note of the many host-government requirements relevant to the transfer of technology via some of these categories of contract. To cite one example, take Iraq. By last report, no Iraqi contracts could be awarded to foreign consulting, engineering, architectural, or construction firms not having established branches in Iraq, unless a branch office was established in Iraq within three months after award of the contract, or the foreign firm arranged a joint activity with an Iraqi counterpart in a partnership agreement. Also, 50 percent of the engineers working for a foreign consulting firm established in Iraq were required to be Iraqi nationals.[3]

Years ago, the late Wolfgang Friedman observed that the contractual joint venture was "much closer to the joint venture as known in U.S. law." This law has been defined, Friedman explained, "as a special combination of one or more persons (partnerships or corporations), where, in some specific venture, a profit is jointly sought without any actual partnership or corporate designation." He went on to point out that "the basic feature of an international contractual joint venture is that investment and expenses are divided between the partners according to fixed percentages."[4] In other words, a contractual joint venture need not have any separate organizational identity; the partners simply enter into an agreement to cooperate in achieving a defined purpose for a specified period of time.

An equity joint venture, on the other hand, stands as a separate, taxable corporation under the laws of the country in which it is incorporated, thereby enjoying limited liability vis-a-vis the claims of creditors (though not necessarily *public* claims arising out of injuries caused by product, toxic waste, and environmental impact), the right to sue, and the capacity to enter into contracts with third parties.

PARTNERSHIPS AND STRATEGIC ALLIANCES

For a long-term, high-risk development type of collaborative activity, which may be ill-defined, a partnership may be considered desirable. Partnerships have become increasingly popular internationally for the development of new technology and the marketing of new products, often going under the name of "consortium," "collaborative agreement," or "strategic alliance." A highly schematized rendering of a possible international partnership is presented in Figure 11–2.

Under this international partnership arrangement, the principal partners retain cost control and access to product marketing. Suppliers provide goods and services to the principal partners. At the department level, each principal partner exchanges personnel and collaborates on the arrangement. The partners delegate authority, personnel, and financial control to a joint management team that actually carries out the terms of the agreement. The joint management team may set up a marketing entity for the partnership; this entity has legal status in both partners' countries, and possibly in other countries. Revenue from the partnership flows through the management team, back to the partners via terms of the partnership agreement.

Twenty-five years ago, Friedman observed:

> It is unlikely that the contractual joint venture will altogether displace the equity joint venture in the constantly changing and evolving relations between capital exporting and capital importing countries. But it does add a new important element of flexibility to the various types of relationships which, depending on the state of economic development, the commodity or utility in question, the political climate, the social and legal systems of the

SELECTING THE TRANSFER MODE / 171

Figure 11-2. International Partnership (or Strategic Alliance).

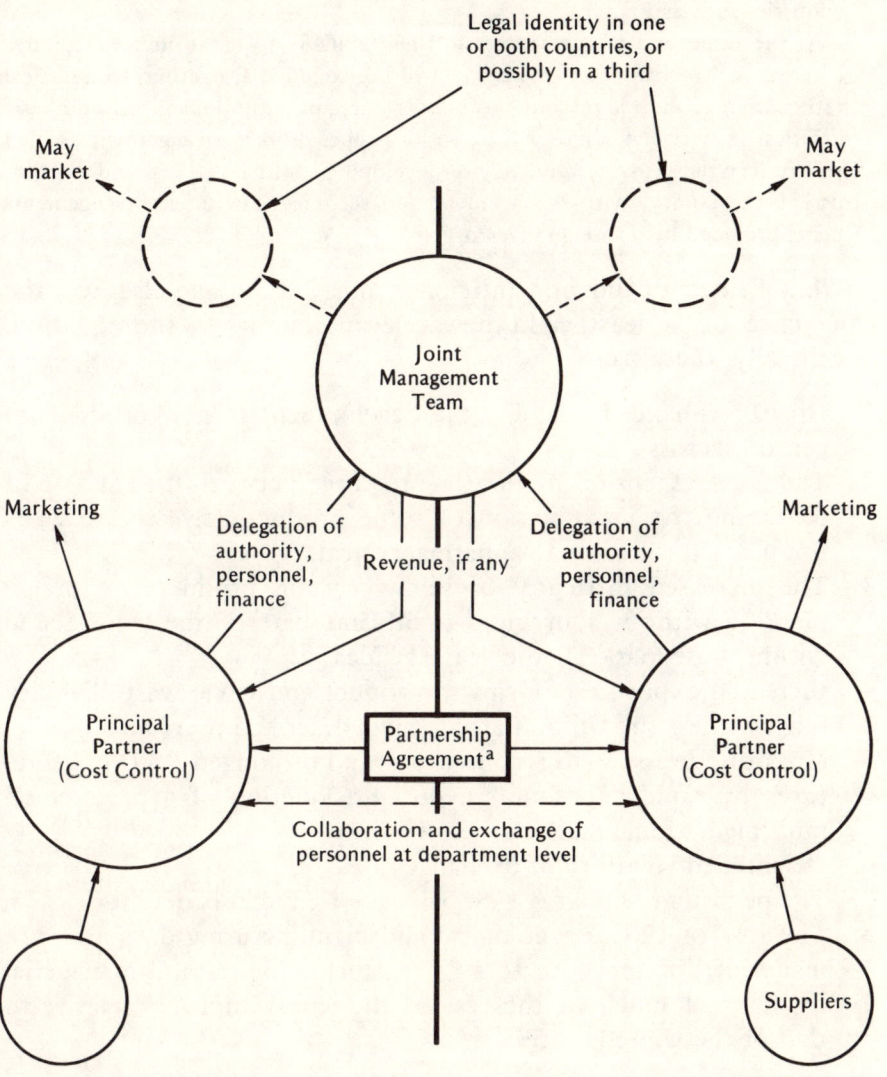

a. Defines purpose, management, duration, division of work, division of revenues (if and when received), a renegotiation process, liability, access to the other firm, investment.

host country, and many other factors, will determine the legal instrumentality to be chosen. The wholly-owned foreign subsidiary will no doubt become increasingly rare, as it is felt to be incompatible with minimum national control and planning by the developing countries—and indeed by developed countries such as France.

At the other end of the spectrum, there is likely to be an increasing range of areas in which foreign ownership will be excluded altogether, for political rather than economic reasons—notably the area of natural resources and basic utilities. In between, there will be various types of joint arrangements, which will reflect the growing tendency of developing countries to demand a majority of the equity, with management and technical assistance arrangements being provided by a foreign investor.[5]

What Friedman did not anticipate, nor did anyone else, was the emergence of at least eight new relevant factors in the equation. Specifically, these are:

1. Greatly enhanced cost for the development of new products in certain sectors
2. The great extension of the time required between the start-up of R&D and the introduction of some products (five to ten years for aircraft, the same for pharmaceuticals)
3. The increased difficulty of selling certain products in foreign markets without sourcing a significant part of the value-added within those markets (the "offset" idea)
4. Increased exposure of firms to product and toxic waste liability (e.g., witness the Union Carbide mess described previously)
5. Mounting pressure to recoup heavy and prolonged R&D expenditures by rapid and simultaneous introduction of a product on multiple national markets
6. Heightened volatility of exchange rates
7. The persistent and large U.S. balance-of-payments deficits
8. The loss of U.S. technological and manufacturing dominance in many—if not most—sectors as the technological and managerial capacity of much of the rest of the world improves relative to that of the United States

These factors have had the effect of creating enormous financial burdens on firms which would go it alone; greatly enhanced risk of commercial failure and of legal attack; a compelling need for a firm to gain rapid visibility and acceptance of products on all major markets; and intensified pressure for the firm to share risk and to participate in technological and managerial developments worldwide.

One could expand the list, but the point is that the need for new forms of international business linkages appeared. These new forms make possible a sharing of financial burdens and risks of various types, limiting the exposure of a firm's own assets and leading to a relatively efficient transfer of technical and managerial knowledge and skills in all directions. Other firms contribute links in the value-added chain, from concept initiation to product commercialization. Efficiency of transfer creates the need for some form of joint management, that is, a continuing and intimate relationship.

One such new form to appear is France's *groupement d'interest economique* (GIE), an example of which is Airbus, a joint French-British-German-Spanish project which has apparently achieved a measure of success even in the U.S. market.[6] Operating under French law, the Airbus GIE, by contract with the four participating firms, exercises management control and performs the marketing function. Engineering and manufacturing are assigned to the member companies. Capital contributions are 38 percent French, 38 percent German, 20 percent British, and 4 percent Spanish. Work shares are roughly the same, though varying somewhat by project.

Modelled after the French GIE is the European Community's economic interest grouping (EEIG), which is a legal entity operating under Community rules rather than national law. (See Figure 11-3.)

After initial registration in one member state, the EEIG is legally permitted to operate anywhere in the Community. In essence, an EEIG is a joint headquarters set up by simple contract which vests collective governing power in the members. An EEIG must have a manager, but no more than 500 employees. There is no minimum capital requirement, and members may make contributions in any form they decide. The EEIG is a vehicle for the promotion of the interests of all participants, but not as a profit-generating entity in that revenues—after deducting the EEIG costs—flow through to the participants and are subject to separate taxation. All members must be actively engaged in some economic activity, must be considered legal residents of the Community, and are jointly liable for any debts incurred unless otherwise specified.[7] What we have here is an international partnership. Certainly, it is neither an equity-based nor contractually based joint venture as previously defined. Because the

Figure 11-3. European Economic Interest Grouping (EEIG), July 1985.

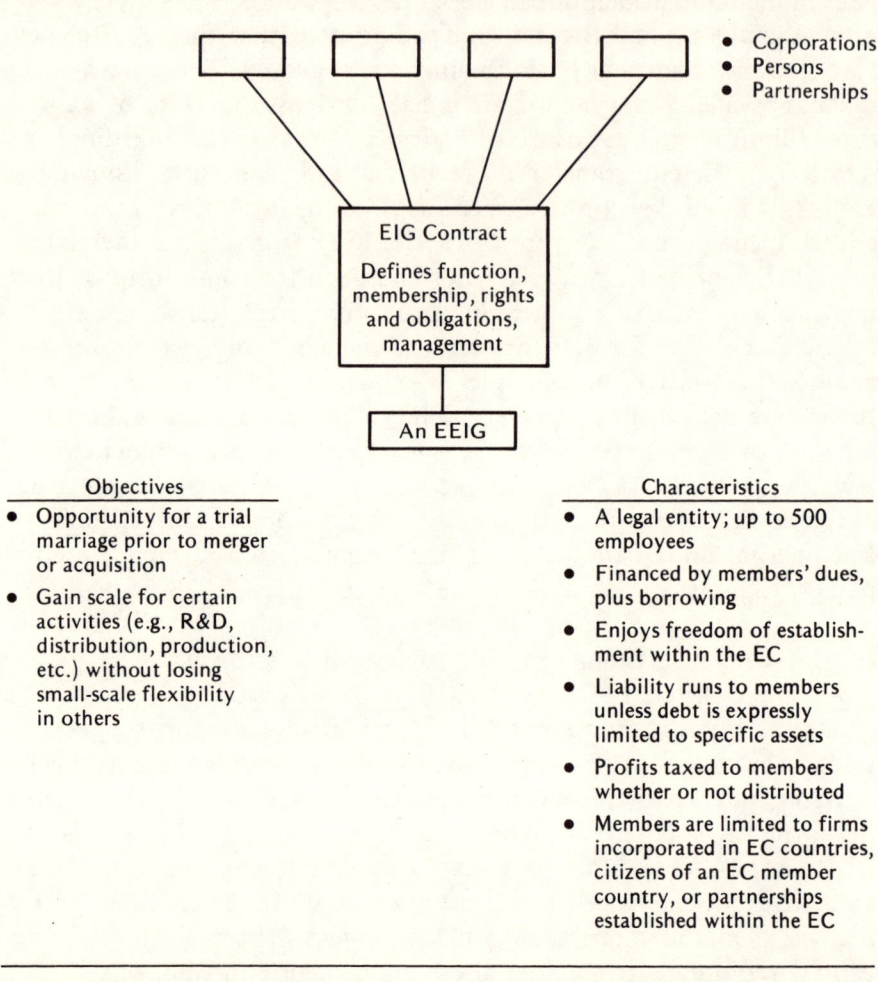

association implies a long-term commitment, it becomes a *strategic alliance*.

STRATEGIC ALLIANCES

One scholar has defined the phrase, "new forms of investment," as including two categories:

1. Joint international business ventures in which foreign-held equity does not exceed 50 percent

2. Various international contractual arrangements involving at least an element of investment from the foreign firm's viewpoint (i.e., a delayed return), but which might involve no equity participation by that firm whatsoever, as is frequently the case with licensing agreements, management, service and production-sharing contracts, and occasionally with subcontracting and turnkey operations.[8]

It was suggested that the underlying cause of the shift by firms to these new forms lay in the inherent conflict between foreign owners and host-country elites. The author added that "whereas foreign investors traditionally felt that ownership was necessary to ensure effective control, the new forms of invesment offer the distinct possibility of a separation of ownership and control."[9]

I would argue the point in that assets located on the other side of an international border from the owners are *never* subject to the same degree of control by the owners as in the purely domestic case, even if those assets are 100 percent–owned. Different laws, regulations, customs, tastes, monetary systems, and distances intervene to limit a firm's effective external control over foreign assets. My own past study in this area has convinced me that governments, both those of the lesser and more developed countries, are becoming increasingly expert at diverting, blunting, and shaping external control over local assets so as to be compatible with local conditions.[10] One effect of such pressure has been the appearance of international partnerships, which Oman, in his otherwise excellent work, fails to discuss.

A 1986 study reported that 60 percent of recently formed international business relationships has not involved the purchase of equity.[11] For example, it was noted, "National Semiconductor is selling Hitachi computers, RCA is selling Hitachi PBX equipment and Matsushita's VCR's, and Honeywell is selling Ericsson telephone switching gear. This new phenomenon is changing how companies will compete in the future." Several reasons were cited for this competitive shift:

1. The rapid pace of technological change and the high cost of development, hence the need to share risk
2. The need for rapid entry of a firm into all markets worldwide
3. The desire to buy supplies and manufactured components at the lowest cost, no matter where produced

4. The end of the one-sided dominance by the United States of the world economy and, in consequence, the growing need for U.S. firms to catch up in respect to manufacturing skills and gaining access to new products developed by others[12]

In a Harvard University study, it was noted that 70 percent of the alliances entered into by American consumer electronics companies with Japanese counterparts involved mutual commitments to distribute and sell each other's products.[13] Many included commitments for joint development. For example, the NMB Semiconductor Company, an affiliate of the Japanese Minebea Company, had tied up with National Semiconductor of the United States for the joint development of very large-scale integrated circuits. A five-year agreement entered into by the two firms called for joint development at a Japanese plant. Subsequently, the U.S. firm was to introduce the jointly developed products into its worldwide marketing network.[14] The list could be extended.

The *Japan Economic Journal* is a rich source for details of these alliances. Headlines tell the story: "Hitachi and GM Will Set Up Joint R&D, Production Projects in 5 High-Tech Areas,"[15] "Japanese, U.S. and British Companies Plan to Set Up an International Telecom Firm,"[16] "Toshiba Will Link Up with IBM Japan to Market General Purpose Computers,"[17] "GE-Fanuc Joint Venture to Research and Develop Factory Automation Systems,"[18] "Measurement Equipment Firms Agree on Extensive Cooperation—Kett and Trebor to Swap Products, Technology,"[19] "More Firms Agree to Foreign Exchanges."[20] This latter reference had to do with the increasing number of Japanese firms stepping up international exchanges of R&D scientists and engineers.

Two far-ranging alliances formed recently between U.S. firms and European are the AT&T-Olivetti relationship, and General Electric and Societé National d'Etude de Construction de Moteurs d'Aviation S.A. (SNECMA).

The AT&T-Olivetti alliance was made public initially in December 1983, when it became known that AT&T would buy a 25 percent equity interest in Olivetti, with an option to increase its position to 40 percent within five years. A master agreement described the motivations leading to the alliance and the management structure to implement the alliance. Olivetti was to have exclusive Euopean distribution rights for all AT&T products related to the office automation market, and AT&T would distribute certain Olivetti products in the United States. In addition, the agreement called for joint activities in the development of new products, with provisions for possible reciprocal production licenses between the two companies. AT&T was apparently pulled into the alliance by the fact that Olivetti offered access to the European market, modern and efficient manufacturing facilities in Europe, and a management which had demonstrated its entrepreneurial capacity through a dramatic turnaround of Olivetti's performance. Olivetti was motivated by a need for certain technology, a stronger presence in the U.S. market, and financial support.[21]

The GE-SNECMA collaboration, signed in 1985 to develop jointly the unducted-fan aircraft engine, has taken the form of two jointly incorporated management companies: one U.S. and one French. Both are staffed by GE and SNECMA, and each has branches in the United States and France, but are run as a single entity, CFM International. These joint firms and their respective branches are responsible for marketing, sales, and product support, thereby providing a single interface for the customer. Each may buy from either GE or SNECMA. Design and component manufacturing responsibilities are split: the core engine to GE, and the fan, turbine, and accessories to SNECMA. There are dual production lines in France and the United States, and the two firms are responsible for the operation of their own facilities, plus related tooling, manufacturing, and inventory costs. The revenues flowing into CFM International, less its operating costs, are split 50-50 by GE and SNECMA. The alleged success of the operation is based on the facts that (1) CFM is an independent entity providing a single interface with the customer, (2) there has been a clear-cut division of responsibilities, (3) both sides gained, (4) there was no overlap of product lines at the start, and (5) both firms are in it for the long haul.[22]

Recurring characteristics of these strategic alliance type of relationships are: (1) little or no direct joint investment by the partners, (2) some form of joint management of certain functions, (3) exchange of personnel and intensified inter-firm communication, and (4) long-term commitments. *One might well look at these partnerships as an attempt by the partners to reduce the cost of external transactions by internalizing them within a long-standing and intimate relationship—a strategic alliance of a partnership form.* Many of the agreements appear to be ill-defined; they are more in the nature of statements of intent to share work, risk, and earnings in respect to a general line of activity or projects over a long period. Such relationships are not subject to annual bidding competition. Integration of any complex system requires close collaboration over a very long period of time. Everything must fit and perform precisely to specifications, whether it be a vehicle, a computer, or an airplane. Over time, it may even become difficult to define a corporation's boundaries. Individual employees may find themselves equally at home in either partner firm. Corporate cultures may converge, and personal loyalties become ambiguous at best. Ultimately, one may be working for the partnership, not for either firm, although he or she may remain on the payroll of one or the other. A classic example may be Boeing's growing inter-penetration of its Japanese associates.

Essentially, Boeing has two U.S.-Japanese partnerships, first with Japanese Commercial Airplane Company, and secondly with the Japan Aircraft Development Corporation. These alliances followed many years of Boeing's subcontracting with Japanese firms for the manufacturing of specific parts. Details of the Boeing alliance, to the extent known, have been published elsewhere.[23] Suffice it to say here that the agreement generally provides that member companies will undertake mutually agreed upon activities, finance those activities, and receive revenue on a pro-rata basis when sales are made. The agreement presumably defines the overall purpose, duration, division of work, division of revenues, a renegotiation process, division of liability, access to the other firm, the exchange of technology and know-how, marketing areas, delegation of management authority to a joint management team, and the division of revenue (if and when received by either partner). No profit is guaranteed either partner in that each controls its own costs.

Apparently Boeing was brought into this new relationship in part because of past experience of one or more Boeing executives with NATO consortia, of which the Boeing alliance is reminiscent. Some of the characteristics of the NATO projects have been national pressure to share in the development process of defense technology; competitive bidding by international consortia of firms with complementary technology and skills (with one firm taking the lead role); establishment of a joint management team (sometimes incorporated as a jointly owned entity (as in the case of the MDDT Corporation of the Martin Marietta lead team for the Multiple Launch Rocket System [MRLS-3]);[24] an overall management agreement; a commitment to assign work roughly equal to the financial commitment of each; and

Figure 11-4. NADGE Project, Early Warning, Tracking and Interceptor System.

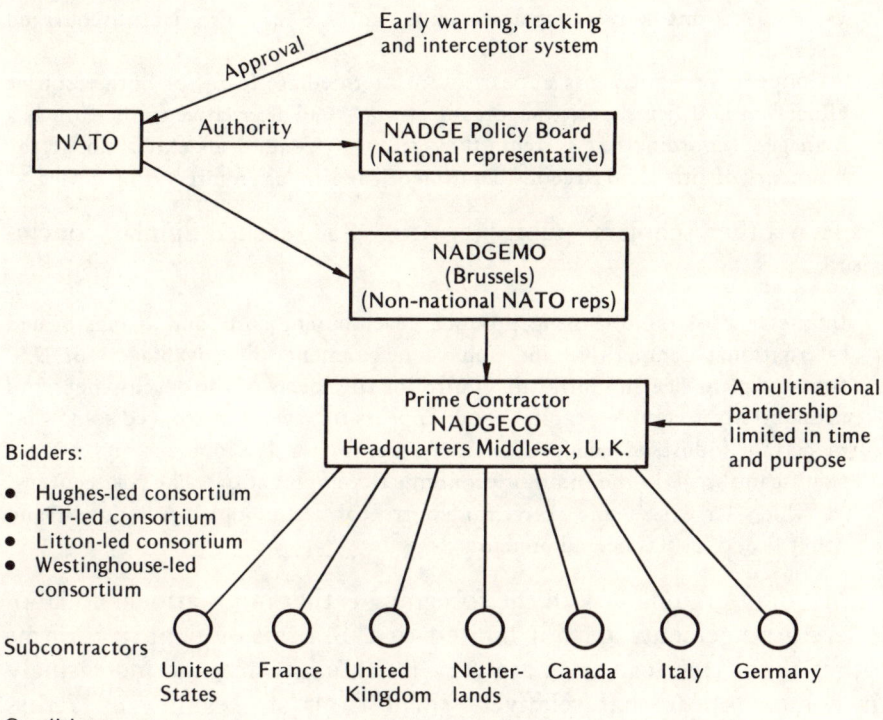

focus on one externally financed project defined by an external body, NATO. These projects and their associated problems have been well discussed in detail by others.[25] (See Figure 11-4 for an example.) These consortia are in the nature of international partnerships, but not of the exclusively commercial type in that they enjoy the commitment of external government financing.

A very perceptive 1983 article which dealt with joint ventures and bidding consortia, concluded in part with these comments:

> The field of study of cooperative agreements between firms seems a very promising one. It has been neglected in the literature in the past, with the possible exception of research on joint ventures.... One reason for such lack

of attention may be traced to a tendency to look at industrial organizations in terms of a sharp dichotomy between markets and individual transactions on one side, and conscious planning within the firm on the other; the great variety of forms cooperative agreements can take may also have discouraged research.

Cooperative agreements appear as an intermediate form for both resource allocation and firms' activities organization. As an alternative to the firm as a managed (coordinated) system, they seem to require an extension of the boundary of firms and a reconsideration of the firm as an integrating device.[26]

Two other scholars, publishing in 1985, reached similar conclusions:

In the face of rapidly rising product development costs and a transformed international competitive and policy environment, the advantages for U.S. firms of giant size and intrafirm control of all aspects of the development and production of complex capital goods appears to have been reduced somewhat in certain industries and products. If true, such a development represents a significant break in the trend of economic development over the past century, in which large size and intrafirm control of technology development and other functions has been dominant.[27]

My only argument with the foregoing is that international collaborative arrangements are not limited to U.S. firms of giant size. Firms other than American ones are very much in the act, and increasingly it would appear that relatively small firms are teaming up across international borders. It may be that the latter have more to gain from such than the giant firms.

NOTES

1. Newly industrializing countries—generally considered to consist of Hong Kong, South Korea, Singapore, Taiwan, Brazil, Mexico, Greece, Portugal, Spain, and Yugoslavia.
2. Irving Jaffe, "The NICs Climb UP the Industrial Ladder: Challenge and Opportunity for the Developed Countries," *The OECD Observer*, no. 147, August/September 1987, p. 12.
3. Information on the relevant requirements, country-by-country, are published periodically by the U.S. Department of Commerce, Domestic and International Business Administration, in the form of *Engineers' Overseas Handbooks*.
4. Wolfgang G. Friedman, "The Contractual Joint Venture," *Columbia Journal of World Business*, vol. vii, no. 1, January/February 1972, p. 56.
5. *Ibid.*, p. 63.

6. *Business Week*, October 20, 1986, p. 34.
7. *Business International*, January 13, 1985, p. 22; *Financial Times*, March 20, 1985, p. 2; "Council Regulation (EEC) No. 213/85 of 25 July 1985 on the European Economic Interest Grouping (EEIG)," *Official Journal of the European Communities*, no. L199/1, July 31, 1985.
8. Charles Oman, *New Forms of International Investment: Developing Countries* (Paris: Organizational for Economic Cooperation and Development, 1984), p. 11.
9. *Ibid.*, p. 12.
10. See R. D. Robinson, *National Control of Foreign Business Entry: A Survey of Fifteen Countries* (New York: Praeger, 1976); *Foreign Investment in the Third World, A Comparative Study of Selected Developing Country Investment Promotion Programs* (Washington: Chamber of Commerce of the United States, 1980); and *Performance Requirements for Foreign Business: U.S. Management Response* (New York: Praeger, 1983).
11. Referred to by Lewis Young, chairman of a study group on international corporate alliances for the Council on Foreign Relations. See "The Corporate Links Abroad," *New York Times*, August 6, 1986, p. 26. Reference in the *New York Times* article is to a study by Albert Bressand, Director, Promethee, Paris, entitled, "Joint Ventures and Inter-Company Alliances in High Technology Sectors" (Brussels, January 15, 1985, manuscript only), which was based on a survey of 974 agreements between companies in industrialized countries between 1982 and the first quarter of 1984.
12. Lewis, *op. cit.*
13. Robert B. Reich and Eric V. Mankin, cited by Lewis, *supra*. n. 10.
14. *Japan Economic Journal*, October 4, 1986, p. 11.
15. *Ibid.*, February 15, 1986, p. 1.
16. *Ibid.*, June 21, 1986, p. 1.
17. *Ibid.*, July 26, 1986, p. 1.
18. *Ibid.*, July 28, 1986, p. 8.
19. *Ibid.*, May 2, 1986, p. 13.
20. *Ibid.*, February 15, 1986, p. 10.
21. Gayle L. Ruedi, "The AT&T-Olivetti Alliance: A Strategic Key to the European Information Movement and Management Market," unpublished paper, Sloan School of Management, Massachusetts Institute of Technology, April 1986.
22. D. M. Hooper, "Collaborative Efforts Between General Electric and Societé National D'Etude de Construction de Moteurs d'Aviation S.A.," unpublished paper, Sloan School of Management, Massachusetts Institute of Technology, March 1986.
23. See Richard Moxon and J. Michael Geringer, "Multinational Ventures in the Commercial Aircraft Industry," *Columbia Journal of World Business* vol. xx, no. 2, Summer 1985, pp. 55–62; and David C. Mowery and Nathan Rosenberg, "Commercial Aircraft: Cooperation and Competition

Between the U.S. and Japan," *California Management Review*, vol. xxvii, no. 4, Summer 1985, pp. 70–92.
24. Subject of a brief study by John N. Brupbacher, "Marketing Defense Technology Abroad," unpublished paper, Sloan School of Management, Massachusetts Institute of Technology, 1985.
25. Suffice to mention two: Jon McLin, "Rationalizing Defense Production in NATO," Parts 1 and 2, nos. 1 and 2, *American Universities Field Staff Reports*, West Europe series, vol. iii (Hanover, N.H., 1986); and Jack N. Behrman, *Multinational Production Consortia: Lessons from NATO Experience* (Washington: Department of State Publication 8593, August 1971).
26. P. Marti and R. H. Smiley, "Cooperative Agreements and the Organization of Industry," *The Journal of Industrial Economics*, vol. xxxi, no. 4, June 1983, p. 70.
27. Mowery and Rosenberg, *op. cit.*, p. 70.

CHAPTER 12

Relating Control and Success in Technology Transfer

As indicated in the previous chapter, joint enterprises may be based either on local corporate law (the equity joint venture) or on a negotiated contract (the contractual joint venture or technical collaboration agreements of various sorts). In either case, one partner may be in a dominant position relative to control (a *dominated joint venture*, or subsidiary); the partners may have roughly equal control (a *shared joint venture*); each may be assigned specific differentiated areas of control (a *divided joint venture*); or neither may exercise any real control (an *independent joint venture*).

CONTROL AND SUCCESS

It should be noted that majority ownership need not be equated with a dominated joint venture. Indeed, a thesis of this chapter is that ownership and control need bear no relationship, and often do not. One early study observed:

> In numerous cases, the foreign partners who participate in joint ventures on a minority basis find that their interests are adequately represented despite the fact that they do not exercise a majority of the voting power in the enterprise. De facto control ... by a foreign partner may be the result of a formal technical assistance or other type of agreement, or it may exist without formal agreement by virtue of the ability of the foreign partner to be of some

special assistance to the enterprise as, for example, in obtaining loans from foreign or international lending agencies. In other instances, there may be business transactions between the joint venture enterprise and the enterprise of the foreign partner which yield such decided advantages to the joint venture that the influence of the foreign partner in it is greater than would be expected from its minority financial participation alone.[1]

On the other hand, it is noted:

> There is a noticeable tendency in joint ventures to allocate managerial functions to the local partners, even if the foreign partners have a majority financial control.[2]

Another early study of international joint ventures reported that executives appeared to equate majority ownership with effective control. "This," it was observed, "could be a rather short-sighted view. A foreign majority position in an operation is immediately open to accusations of foreign dominance in a local firm. As a result, it becomes more 'visible' to local countervailing forces. Because of the existence of such forces, it is not necessarily safer in the long-run than would be a foreign minority shareholding [or a purely contractual relationship]. In the short run, these forces may also work to restrict a foreign majority's freedom of operation."[3] The reference here, of course, is to the imposition of political restraints in the form of performance requirements.

In fact, strategic and/or operational control in a joint enterprise may be shared jointly via the board of directors or a managing board on which both partners have equal votes, or at least a veto on specific issues. If local corporate law permits, such representation or voting rights need not rest on the proportion of board seats held or percentage of ownership. Areas and degrees of control may be defined separately by special provisions in the articles of incorporation or corporate by-laws, or by special contract, although again local corporate law may define the latitude enjoyed in this regard by requiring that control in *certain* areas be linked to degree of ownership. In any event, a continuum of control by a partner is theoretically possible, from complete control (strategic and operational), to operational control only over specified functions, to complete silence (i.e., nonparticipation). Figure 12–1 diagrams the various types.

One scholarly study suggests that these various types of joint enterprises should be used according to the relevant managerial skills and knowledge each partner can bring to the joint effort, as diagramed

Figure 12-1. Differentiating Joint-Enterprise Type.

		Firm A's Relevant Managerial Skills, Technology, and Knowledge		
		None	Partial	Complete
Firm B's Relevant Managerial Skills, Technology, and Knowledge	None	No joint enterprise	No J-E or independent J-E	Dominated J-E (by Firm A)
	Partial	No J-E or independent J-E	Divided J-E	Shared strategic management if necessary, but Firm A dominates operations
	Complete	Dominated J-E (by Firm B)	Shared strategic management, if necessary, but Firm B dominates operations	Shared strategic management only; either partner dominates operations

Source: Adapted by the author from Table 4:1, in Peter Killing, *Strategies for Joint Venture Success* (London: Creom Helm, 1983), p. 55. (Note that the table in Killing was incorrectly drawn.)

with some modification in Figure 12-2. The study claims that "the dominant [i.e., dominated] . . . ventures have a much better success rate than shared management ventures." It is also claimed that "shared management ventures should not be established unless it is abundantly clear that the extra benefit of having two parents managerially involved will more than offset the extra difficulty which will result." But the author goes on to admit that even though shared management ventures are the most difficult to manage, "in some cases they are unavoidable."[4] By not sharing management, an even higher cost is incurred, such as not getting access to least-cost resources, including technology, either because a partner feels it necessary to extract a higher price for its contributions in the absence of effective control or the local parent government refuses certain services or benefits (e.g., access to local capital, markets, tax exemptions, etc.).

Figure 12-1, of course, is meant only as a guide. Even though one or both partners possess *all* the relevant management skills, technology, and knowledge needed, one or both may see the venture as rela-

186 / SPECIAL ISSUES

Figure 12-2. Selecting the Joint-Enterprise Type.

Control \ Ownership	Equity Joint Venture (divided equity participation)			Contractual Joint Venture, Partnership, or Technical Collaboration Agreement (no equity participation)
	Majority-owned by one partner	Equally owned[a]	Owned in any percentage (i.e., variable)	
Dominated management (subsidiary)[a]				As defined and assigned by contract, but with one partner being dominant at least in respect to operational management
Shared management[a]				As defined and assigned by contract, with rough equality in strategic and operational management
Divided management[b]				As defined and assigned to the partners by contract, with virtually no control residing in an *overall* management body or general manager
Independent management (can be looked upon almost as though it were a business trust)[c]				As defined and assigned by contract to a virtually autonomous management

a. Some areas, and degree, of control may be separately defined and assigned to one of the partners either by contract or by special provisions in the joint venture's articles of incorporation and/or by its by-laws, residual control resting with the board of directors or some other body.

b. Virtually all areas, and degree, of control *must* be defined and assigned to one of the partners either by contract, the articles of incorporation, or the by-laws, with very little overall control residing in the board of directors of the joint venture.

c. Technically, a business trust is an organization in which one or more trustees manage assets for specific purposes, assets in which others have a beneficiary interest but no control other than that specified in the trust deed (i.e., contract) and in law. Ownership may be evidenced by shares.

Source: Compiled by Richard D. Robinson.

tively insignificant to its total operations and, hence, be willing—or even anxious—to allocate operational control to the other in a dominated joint enterprise, or to participate in setting up a virtually independent joint enterprise simply to provide some goods and services, with the production of which neither parent wishes to be closely involved. In any event, as this same scholar points out, using a guide such as Figure 12-1 is "complicated by the fact that the importance of a partner's managerial [or technical] contributions will probably change over time."[5]

It has been observed that the degree of independence enjoyed by a joint enterprise manager is at least in part a function of performance. Low performance tends to lead to intervention by one or both partners, which in turn reduces efficiency of decision-making in the joint venture, further reduces performance, and thereby leads to even greater partner intervention.[6] The contrary argument might likewise be made that a highly profitable joint enterprise could be perceived as so important to overall corporate profit by one or both partners that very close monitoring of the joint venture management is deemed desirable. It is my own view that a truly independent joint enterprise is a very rare animal unless it is so small and unimportant that neither partner is greatly concerned so long as it performs its assigned function reasonably well.

Success of joint enterprises should not necessarily be equated with long life. If major objectives of one or both partners are achieved relatively quickly in terms of effective transfer of technology and skills, establishment of a market niche, or the building of a local or international distribution system, the fact that the joint enterprise relationship was transferred into either a parent-subsidiary relationship (in which unambiguous control is assumed by one of the partners), merged with another entity owned by one of the former partners, or "downgraded" into a purely contractual relationship (e.g., license, technical, management, marketing, and/or supply) may be quite satisfactory to both parties. The transformation should not necessarily be considered a failure. Flexibility is the hallmark of successful international business relations in today's world of volatile exchange rates, global markets, and rapidly shifting economic and political conditions. Only if the venture results in commercial failure, loss of assets, mutual recrimination, and no continuing, mutually beneficial relationship can it really be dubbed a failure. Insofar as is known, the failure rate of joint ventures, partnerships, and technical collaboration agreements in these terms has not been the subject of study.

The question of success really comes down to the ability of a country (or enterprise) to capture the foreign savings, foreign technology and skills, and the domestic and/or foreign market share that it desires, and at acceptable cost. Entering into an international joint enterprise—equity or contractual joint venture, partnership, or technical collaboration—is a way for doing so.

Quite apart from a compatible legal-political environment, certain managerial characteristics appear to be associated with joint-enterprise success as defined above. An early study lists seven of these factors.[7] I list nine.

1. Selection of an appropriate partner
2. Frank statement and discussion between the prospective partners of their respective goals, and adjustment of any goal incongruities which emerge so as to be tolerably compatible
3. Explicit definition of areas of control required by each partner and the means of assuring such control
4. Explicit evaluation by each partner of the costs and benefits accruing to it from participation in the venture, and adjusting the perceived values of these flows so as to be agreeable to each partner (statement of the basis for various prices)
5. Creation of a mechanism for ongoing adjustment as those flows and their perceived values change over time
6. Explicit discussion of, and agreement relative to, problems having to do with the staffing, loyalty, and management of the joint venture (including such management processes as planning, financial reporting, monitoring of performance, management information systems, product testing, industrial relations, etc.)
7. Explicit agreement as to the conditions for termination, whether by liquidation, sale, or takeover of the joint enterprise by one partner, and/or transformation of it into a set of contractual relationships
8. Explicit agreement that issues incapable of resolution by internal negotiation be submitted for arbitration in a third country according to a specified process

9. Establishment of a system that facilitates unambiguous separation of joint-venture costs and revenues from those of either partner

I deal with each of these characteristics in some detail below.

Selecting an Appropriate Partner

It is patently obvious that in selecting potential partners in a joint enterprise, a firm should seek one with a known track record in respect to credit-worthiness, integrity, dependability, and managerial competence. By careful study and evaluation, it is possible to obtain fairly good readings on these attributes. Financial records, reputation in the marketplace, performance of contractual obligations, and managerial effectiveness can be measured at least approximately.

Experience suggests that structure, managerial style, function, size, prior joint-enterprise experience, and the geographical diversification of prospective partners can be important factors in enhancing an enterprise's success. That is, the more nearly equal the two prospective partners are along these dimensions, the greater the likelihood of a successful enterprise. By *structure*, one refers to ownership (e.g., a publicly held vorporation *vs.* a closely held or family corporation), divisional setup (uni- or multidivisional, and the basis for divisionalization—product, region, function), numbers of layers of management, etc. *Managerial style* refers to the nature of superior-subordinate relations in decision-making, continuity of managerial style as one goes down the hierarchy, the formality of communications, and the degree of intuition vs. objective analysis used in decision-making.[8] *Function* refers to a firm's principal activity (e.g., manufacturing, sales, banking, trading, etc.). As for *size*, one scholar observed that "a significant size mismatch between a venture's parents can create a lot of problems for the venture."[9] The point is, of course, that the venture is likely to be seen as much more important by one partner than by the other. The final two characteristics—*prior joint-enterprise experience* and *geographical diversification*—are self-explanatory. The partners would, as a result of similarity along these two dimensions, be at similar points on their respective learning curves.

In essence, we are talking about factors which tend to define the "organizational culture" of the partner firms. The closer these cultures are, the easier it should be for the firms to cooperate in a joint enterprise. In some sense, they speak the same organizational language. It almost goes without saying that an organization's culture is strongly influenced by the culture of the larger society surrounding it. The more similar the cultures of the parent countries, the more similar the cultures of organizations based in those countries are likely to be—all other things being equal, which, of course, they rarely are. One scholar makes a similar but more limited observation by reporting that "the more similar the culture of firms forming a shared management joint venture [or partnership or collaboration], the easier the venture will be to manage." "Culture is considered to have two components," he continues, "One being the culture of the country in which a company is based, one the 'corporate culture' of a particular firm in question."[10] Personally, I would not limit this observation to *shared* joint enterprises, but extend it to all types. In *all* cases, communication between the partners is critically important, and similarity of culture—national and corporate—facilitates that communication.

One study reported that executives did not, in fact, identify cultural differences as a source of serious problems in joint ventures and partnerships. But, the same author went on to conclude, "the managerial philosophy of executives in charge of joint ventures was a very important factor in joint venture longevity and smooth operation." In the sample of joint ventures used in that analysis, all those which had been terminated had experienced a change in management shortly before in at least one of the partner firms, which change, "had resulted in substantial change in the parents' philosophy towards the joint venture."[11]

It is possible that a joint-enterprise decision-making process can generate a deadlock, "if the partners have not developed equitable mechanisms for resolving" such impasses.[12] One author suggests that "it is generally more advantageous to draft a *team* of operating managers that maximize tradeoffs between synergies [from shared resources][13] and economies [from centralized facilities] during the courtship stage. The team should be taken from both owners' personnel and should work together during the 'trial marriage' period."[14] The implication is that if such common benefits cannot be found, one is joint venturing or in partnership with the wrong firm.

At times, complaints have been lodged concerning the limited authority demonstrated by home office executives who had been sent overseas to attend board meetings or carry on substantive problem-solving or negotiating discussions. "In some cases, the board of directors of the . . . [foreign parent] company had subsequently refused to recognize agreements that had been made by home office executives sent overseas to represent the company for the achievement of just such agreements."[15] Obviously, such a situation can provide a most inauspicious launching of a joint undertaking.

Explication of Partners' Goals

A number of studies emphasize the importance of a frank exchange of goals between partners in a proposed enterprise and of an articulation of these, whether in the by-laws or in a separate statement of goals and strategies. A potential source of problems in joint ventures, and presumably in partnerships and technical collaboration agreements as well, one study reported, was "the lack of written records or chronological files in which the relationship between the partners was recorded as it developed." Objectives of the partners may diverge, and the absence of chronological records aggravates such situations. "Without such records, developments often go unnoticed until problems have reached disastrous proportions, by which time, in most cases, it is too late to apply simple remedies. Crises could be brought about by the accumulation of small problems over a period of time. In these cases, either the lack of periodic sessions between the partners in which the small differences were aired, or the unwillingness of individual executives to bring up small problems in these sessions, were responsible for failure."[16]

In listing possible goal incongruities (Figure 12-3) I have in mind the situation involving a relatively large foreign-based corporation with several operations outside its parent country vis-a-vis a smaller, domestic enterprise in a less developed country with none, or few, operations abroad.

Lying behind these possible goal incongruities are, *first*, possibly differing time values of money (i.e., different "liquidity preferences," use of different interest rates, and, hence, different costs of capital); and *second*, differing perceptions of risk due to a "portfolio" or "diversification" effect that is greater for one than for the other. Therefore, long-term planning and investment in somewhat riskier projects may make sense for one, but not the other. A *third* incongruity may arise out of the very fact that the larger foreign partner

Figure 12-3. Possible Goal Incongruities.

Possible Goals Relative to:	For the Larger Foreign Firms	For the Smaller Domestic Firm
Profit	To maximize profit over an internationally integrated system ultimately denominated in the parent-country currency	To maximize profit domestically, ultimately denominated in local currency
Timing of payouts	To maximize returns over a relatively long growth period with little immediate payout	To maximize returns in the relatively short run with immediate payout
Level of payouts	To pay a minimum dividend, or a dividend adjusted to payments from other ventures, thereby "normalizing" total revenue to the partner	To pay maximum dividends to the partner with much less adjustment for other income
Growth	To be achieved via product and/or geographical diversification	To be achieved via increased market share in one of a few markets with limited product diversification
Investment	To invest relatively large amounts in research and development, growth, and product diversification	To invest relatively little in R&D, growth, and product diversification
Market	To target sales principally to the market domestic to the joint enterprise, if the firm has many operations elsewhere, but might be worldwide if the enterprise has a "global product mandate"	Principally export, but might be principally domestic if the firm has no, or limited, experience or expertise relative to foreign markets and an attractive domestic opportunity exists
Technology	To develop new technology and protect it from use by others	To use existing technology without payment of royalty or fee
Production	To specialize in a given function, product, or process, wherever economic advantage for doing so exists (i.e., vertical integration between the joint enterprise and other enterprises in the system)	To maximize local value-added (i.e., vertical integration within the joint enterprise or between the joint enterprise and local partner)
Employment	To shift redundant managerial personnel elsewhere in the system to the joint enterprise	To shift labor redundant in the domestic enterprise of the partner to the joint enterprise
New product or process trial	To use the joint enterprise to test out new processes or products so as to share risk	To use developed processes and products with demonstrable markets

Source: Compiled by Richard D. Robinson.

is looking at multiproduct global markets; the smaller domestic firm, at a single-product one-country market. Revenue variance for the former (i.e., a measure of uncertainty) may be substantially less overall than for the latter, which is relying on sales of fewer products in fewer markets, perhaps even one product in a single market. Furthermore, the large, diversified firm may have a large *internal* market for the joint-venture product, but not the local partner. Hence, the impact of *external* market factors may be more delayed for the former, also leading to less variance in benefit.

Such goal incongruities can only be worked out by laying them on the table at the beginning and devising means of either ameliorating their impact or adjusting for them. For example, it might be useful at the outset to define the joint enterprise as a profit-center (or not, as the case might be), and in so doing carefully specify how all transfer prices between the partners and the joint enterprise are to be established. The point is to reduce the latitude either partner may exercise on transfer prices—either directly or via influence on the joint venture (e.g., the ability to stop the flow of technology or some needed skill or input to the joint venture). Such prices refer to *any* transfer in either direction which may show up as sales or marketing commissions, purchase discounts, license royalties, technology transfer fees, training charges, consulting fees, management fees, overhead burdens, interest rates, reimbursements for local personnel, etc. Whenever possible, initial prices should be established and any change linked to some objectively determined index or mutually agreed upon change in value to the joint enterprise.

Possible bases for setting transfer prices, and their subsequent adjustment:

Possible Bases

1. Arm's-length market price, plus or minus a negotiated discount or premium [e.g., LIBOR (London Inter-Bank Offer Rate) ± a specified percent for interest rate on inter-enterprise loans]
2. Cost, plus a negotiated profit (requires the right to verify via access to the supplying firm's books)
3. Negotiated price, presumably at some point between cost (including "opportunity cost" of a possibly foregone market or alternative use) and economic value (i.e., scarcity value as measured in

terms of contribution to profit), or an independently determined market price, where available

Possible Adjustments

1. Use of SDR- or ECU-denominated prices (Special Drawing Rights or European Currency Unit, both of which are published daily) so as to minimize loss and gain resulting from exchange-rate changes between the currency of the host country (to the joint enterprise) and that of the foreign partner (or the currency in which the transaction is denominated, if it is different)

2. Automatic periodic adjustment of price as agreed upon indices of inflation in the two relevant countries move out of phase with the shifts in the exchange rate. That is:

$$\left[1 + \left(\frac{P_1 - P_0}{P_0} \bigg/ \frac{X_1 - X_0}{X_0}\right)\right] \times \text{base period price of good or service}$$

Where: P_1 = an index of general price level at end of period
P_0 = an index of general price level at beginning of period
X_1 = foreign exchange rate at end of period
X_0 = foreign exchange rate at beginning of period

Definition of Areas of Control Desired

Control comes in two forms: *positive control*, which leads or encourages decisions; and *negative control*, which is equated with the ability to block a decision. Given the contributions a partner firm expects to make to a joint enterprise, and the countervailing stream of anticipated benefits, each partner will wish to exercise some degree of control—whether positive or negative—to both limit the costs incurred and protect the benefits to be enjoyed. It is not useful to speak of control as though it were an unsegregated whole. It is more useful to be explicit at the start about the differential areas of control possible, none of which are dependent upon ownership unless local law—or agreement between the partners—specifies otherwise. A list of the more significant areas of control, and some mechanisms for exercising such, are presented in Figures 12-4 and 12-5.

Figure 12-4. Areas of Control and Some of the Relevant Mechanisms.[a]

Possible Areas of Control	Possible Control Mechanisms
Marketing Strategies	
Choice of goods and services to be marketed	1, 7A, 11
Exclusivity of marketing rights to be granted	1, 7A
Selection of target market(s)	1, 3, 7A, 11
Degree of market penetration to be sought	1, 7A
Choice of Channels to be used	1, 7A
Customer service to be given	7A
Promotion (product differentiation) effort	7A
Pricing	
Sourcing Strategies	
Quality control	3, 4, 7E, 8, 18
Degree of local integration (i.e., local value-added)	7E, 18
Make-or-buy decisions	6, 7B, 18
Choice (and source) of capital equipment purchased and technology transferred	4, 7C, 7D, 9
Organization of production	7E
Production site selection	6, 9, 11
Research and development (location, substance, budget)	7D, 9, 11
Personnel Strategies	
Recruitment	7E
Pay scales and other benefits	6, 7E, 10, 16
Industrial relations	6, 7E, 11
Training and development	7E
Managerial style (previously defined)	7E
Promotion	6, 7E, 10, 16
Termination	6, 7E, 10, 16
Financial Strategies	
Definition of profit center	11, 12
Setting of transfer prices between joint enterprise and partners (or their associated firms)	2, 12, 18, 19
Establishing contractual relations between the joint enterprise and partners (or their associated firms)	6, 11, 19
Financial structure (debt/equity)	6, 11, 12
Establishing a payout policy	6, 11, 12
Specifying sources of financing	6, 11, 12
Covering risk (commercial, casualty, political, foreign exchange)	7E, 9, 12, 13, 14, 15, 19, 20, 21

Figure 12-4. continued

Possible Areas of Control	Possible Control Mechanisms
Legal Strategies	
Specifying the legal relationship between the joint enterprise and the partners	7, 11
Protecting valuable rights (patents, trademarks, copyrights, trade secrets)	3, 4, 5, 7, 17, 21
Control Strategies	
The planning function (e.g., growth, geographical diversification, product diversification)	11
Establishing and maintaining an adequate reporting system (financial, results against planned goals, costs per unit of production, production per man-hour, etc.)	12, 15, 19, 21
Evaluation of performance	11, 13, 14
Public Affairs Strategies	
Public image	19, 20
Political involvement	19, 20
Environmental impacts (product, toxic waste, environmental liability)	8

a. The numbers refer to items on Figure 12-5.
Source: Compiled by Richard D. Robinson.

Figure 12-5. Some Control Mechanisms.

1. Requiring the use of the partner's own distribution channels
2. By specifying the partner's rights to monitor invoices issued, and payments made, by the joint enterprise
3. By retention of trademark ownership by the partner and licensing use to the joint enterprise
4. By retention of patent ownership by the partner and licensing use to the joint enterprise
5. By requiring joint-enterprise employees needing to know unpatented trade secrets to sign non-disclosure agreements as a condition of employment
6. By requiring permission of one or more of the partner's representatives on the board of directors, regardless of share in ownership
7. By separately negotiated (A) marketing agreement, (B) purchase agreement, (C) license agreement, (D) technology transfer agreement, (E) management contract
8. By having a technical representative(s) in the joint-enterprise plant with authority to stop production
9. By limiting the joint enterprise's right to make an investment up to a specified amount within a given period of time without agreement of the partner's representative(s) on the board of directors
10. By requiring that the selection, promotion, and salaries (and other benefits) of all joint-enterprise personnel over a specified level be approved by the partner's representative(s) on the board of directors
11. By requiring that any deviation from the goals and objectives in the initial agreements be approved by the partner's representative(s) on the board of directors
12. By requiring that the partner be given full access to the joint enterprise's books
13. By requiring periodic audit of the joint enterprise's financial position by an independent auditing firm and communication of same to the partner
14. By requiring that reports be submitted to the partner in a form and language stipulated by it
15. By requiring the bonding of any joint-enterprise executive authorized to pay out funds over a specified amount or to enter into contracts with third parties
16. By drawing up a standard employment contract
17. By the partner's retaining critical pieces of the relevant technology
18. By requiring that the joint enterprise purchase components of a design and quality available only from the partner

Figure 12-5. continued

19. By requiring that copies of all correspondence on certain subjects be provided to the partner if so requested
20. By requiring approval by a partner's representative of all joint-venture advertising copy
21. By staffing with the partner's own personnel and either keeping them on the partner's payroll or limiting their assignment to the joint enterprise, clearly indicating that their long-term careers (e.g., promotion, benefits, retirement) are with the partner, and/or keeping them intimately informed of what goes on in the partner firm (via letter, attendance at meetings, etc.)

Source: Compiled by Richard D. Robinson.

It is suggested that the joint-enterprise strategy is most attractive to prospective partners, both foreign and domestic, where the legal environment permits and supports these various control mechanisms by giving the managements of the partner firms sufficient latitude to employ them. It is not suggested that all, or even most, of these mechanisms be imposed on a given joint enterprise at any one time, or perhaps ever. Indeed, as the partner firms learn to live harmoniously with one another in a joint-enterprise relationship, corporate cultures (i.e., ways of doing things and their respective views of the world and of themselves) tend to converge. Individuals come to know one another intimately. Over time, strong mutual interest and trust may develop in both directions, and virtually all controls may well be relaxed—that is, other than those necessary for planning purposes by the partners themselves in their own operations.

One should be aware that controls are not cost-free. Such cost may appear as direct, out-of-pocket expense, or in the form of a slowdown in decision-making, and subsequent missed opportunities. A joint enterprise may become so heavily burdened with partner-imposed controls that it becomes exceedingly high-cost, non-competitive, and eventually, inoperative. On the other hand, an independent joint enterprise may run counter to the interests—and profits—of both partners (e.g., by competing in the markets of one or both partners).

The point is that a foreign firm considering a local joint enterprise—be it equity joint venture, contractual joint venture, partnership, or technical collaboration agreement—is likely to look with disfavor on the joint-enterprise proposal unless assured that it can protect its name, existing markets, technology, know-how, financial assets, scarce personnel resources, ongoing contractual obligations, and its public image. Initially, it is pointless to talk of basing the agreement on mutual trust and confidence; that comes with long association—if ever. At the start, it is exceedingly risky for either partner to rely upon an informal "gentleman's agreement."

It is well to realize that a partner's need to intervene (i.e., control) in decisions made in a joint enterprise increases as:

1. the venture's strategic importance to the partner increases;
2. the value and scope of resources shared with the partner increases; and
3. the degree of resource transfer between partner and joint enterprise (i.e., vertical integration) intensifies.[17]

Explicit Evaluation of Costs and Benefits

It has been demonstrated by experience that, apart from an explication of each partner's goals and the adjustment of goal incongruities so as to be tolerably compatible, an examination of possible costs incurred and benefits to be realized by each partner participating in a joint enterprise should be carefully analyzed. Figures 12-6 and 12-7 specify some of the possible costs and benefits involved.

To the extent possible, each partner should evaluate its own costs and benefits and those of the other partner. In essence, a successful

Figure 12-6. Costs Involved in Participating in a Joint Enterprise.

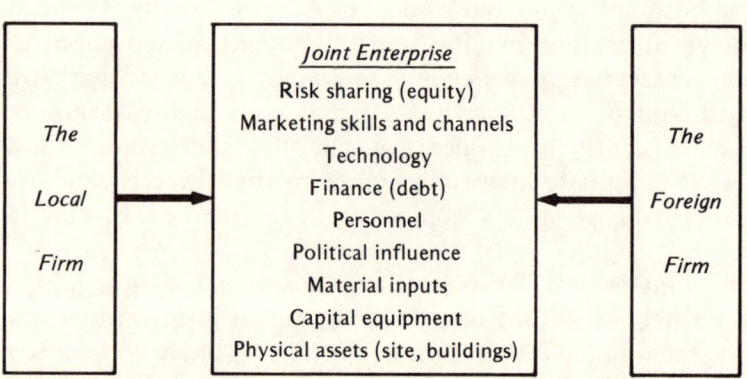

Figure 12-7. Possible Benefits Derived from Participating in a Joint Enterprise.

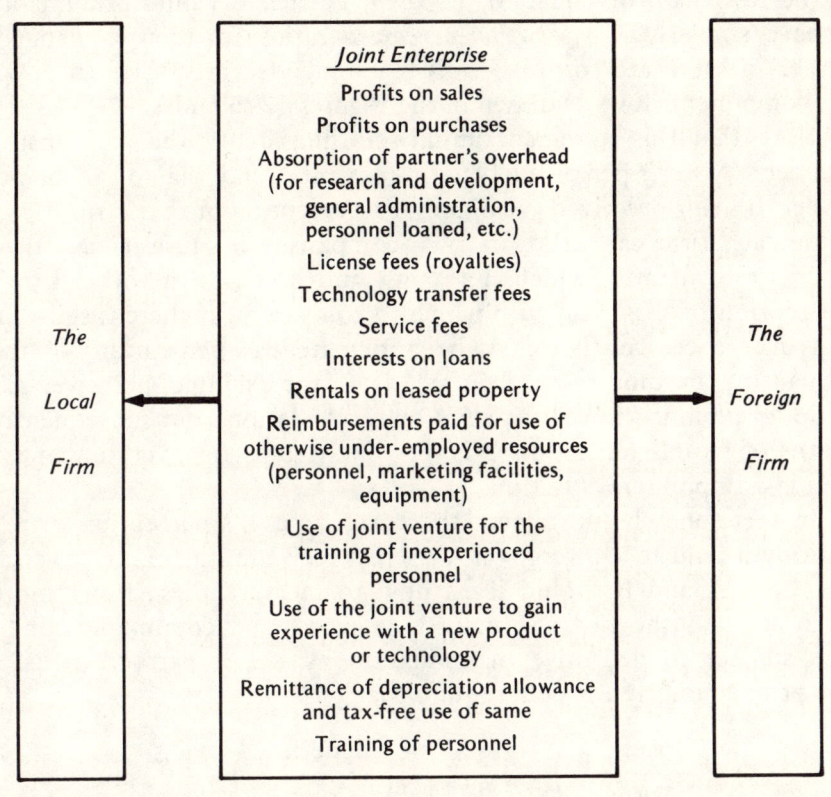

negotiation would then undertake an exchange of the "other" perceived costs and benefits. One can portray what happens with a set of equations.

$$\text{Self-Perceived} \qquad \text{Other-Perceived}$$

$$(1) \quad \left(\frac{\text{Benefits}_A}{\text{Costs}_A}\right)_A \geq \left(\frac{\text{Benefits}_B}{\text{Costs}_B}\right)_A$$

$$(2) \quad \left(\frac{\text{Benefits}_B}{\text{Costs}_B}\right)_B \geq \left(\frac{\text{Benefits}_A}{\text{Costs}_A}\right)_B$$

The first ratio simply indicates an estimate of benefits to Partner A over the costs to Partner A as *perceived by A*. There would be a negotiation only if Partner A perceives Partner B's benefit/cost ratio to be less than, or equal to, its own. Partner B would presumably behave similarly. These *other-perceived* ratios are then exchanged, either in aggregate form or—as is recommended here—broken down by component flows as drawn in the Figures 12-6 and 12-7.[18]

Note that this statement implies nothing about the relationship between A's *self*-perceived benefit/cost ratio and that of B, or between B's *self*-perceived ratio and A's perception of that ratio. However, each firm can make its own comparison because it has calculated its own ratio, which it retains, and now has the other firm's perception of that ratio with which to compare it. If there is substantial difference, conflict may be anticipated as perceptions of the other firm become more accurate. If B, for example, perceives A's ratio to be much lower than A's own calculation, and subsequently learns of its miscalculation, B is very likely to become very unhappy, and insist upon renegotiation.

In fact, one should expect that over time, B's perception of A's ratio will tend to converge on A's self-perceived ratio, and vice versa, for both A and B should learn how to measure these flows more accurately as the flows actually take place, and communication is less subject to deceptive manipulation by either partner. Thus, a long-run equilibrium would require that:

$$(3) \quad \left(\frac{\text{Benefits}_B}{\text{Costs}_B}\right)_A = \left(\frac{\text{Benefits}_B}{\text{Costs}_B}\right)_B \quad \text{and,}$$

$$(4) \quad \left(\frac{\text{Benefits}_A}{\text{Costs}_A}\right)_B = \left(\frac{\text{Benefits}_A}{\text{Costs}_A}\right)_A$$

These relationships imply further that:

$$(5) \quad \left(\frac{\text{Benefits}_A}{\text{Costs}_A}\right)_A = \left(\frac{\text{Benefits}_B}{\text{Costs}_B}\right)_B$$

That is, if the partners' self-perceived benefit/cost ratios differ significantly from one another, conflict of interest can be expected to surface. That conflict will be translated into overt dissatisfaction as one or both of the parties become aware of this disparity even though the initial conditions (equations 1 and 2) were met. Note that there is no mention of *actual* benefit/cost ratios, only of the self-perceived and other-perceived, for reality is only relevant as it is perceived.

One is now in a position to postulate a series of questions. From the point of view of *one* of the partners, these are:

1. What value does the firm (e.g., Firm A) place on what it proposes to contribute to the joint enterprise?
2. What value does it place on Firm B's contribution?
3. What value does A place on the benefits it expects to enjoy?
4. What value does A place on B's benefits?
5. How certain is A of the valuations it places on these flows?
6. What restraints (controls) does A need to provide assurance that the values it placed on these flows will be realized?
7. What restraints (or controls) by B can be anticipated?
8. What is the cost of maintaining those controls or influences seen as necessary by A?
9. *How will the value of all of these flows (and controls) shift over time?*

The other partner (Firm B in this example) would make its own determination of these factors.

A Mechanism for Ongoing Adjustment

The importance of the last query lies in the fact that if any of the values shift significantly, mounting pressure for a compensating shift in the flows should be anticipated and for the tightening or relaxation of controls imposed on the enterprise by the partner firms. By anticipating such shifts, adjustment may be effected with minimum friction and cost—in both psychological and financial terms—and, thereby, greatly enhancing the possibility of success for the joint enterprise.

What is obviously called for—and indeed is the practice in many successful joint enterprises—is either an ongoing or periodic renegotiation in very specific terms between the partners to adjust contributions, benefits, and control mechanisms in the manner suggested above. It is also urged that these negotiations be undertaken at the board level, or perhaps by a specially appointed negotiating committee.

The literature is full of rather loose statements about the need for flexibility in successful joint enterprises. One scholar makes the point that a "joint venture's [enterprise's] need for each of its partners may change over time." He points out that "clearly the learning which takes place ... reduces the parents' need for one another and weakens the ties that hold the joint venture [enterprise] together."[19] I would argue that although ties may be weakened, in many cases they will be *strengthened* as time passes if a continuing negotiation occurs of the sort suggested. A recent study reports that "joint venture [enterprise] pay-offs [i.e., benefits] to their respective parents had been frequently renegotiated."[20] Renegotiation of payoffs makes little sense, however, unless costs (i.e., contributions, in my terminology) had likewise changed or been renegotiated. One study makes the point that, "adaptive [control] systems are needed to provide for the changes that occur in the relationships between joint-venture partners, between owners and ventures, or within the venture's domain."[21] The same would seem to apply to the partnership or collaboration case.

Explicit Agreement on Joint-Enterprise Staffing, Loyalty, and Management

One relevant study concludes, "staffing is clearly an important part of the design of joint ventures." The author's conclusions on joint-venture staffing are two: (1) "dominant parent joint ventures [dominated joint enterprises in my classification] should not employ managers from their passive parents"; and (2) "to enhance information flow and capture the skills of parent companies, shared management ventures between firms from similar cultures should employ managers from both parents."[22]

Presumably, partnerships and technical collaboration would be included as well. This same scholar felt that more research was needed before a conclusion could be reached relative to the staffing of shared-management enterprises between parents of significantly different cultures. He was altogether silent in respect to independent joint enterprises and what I have called "divided joint enterprises." Presumably, in the case of the latter, the function allocated to each partner would be performed by that partner's personnel. And, for an independent joint enterprise to operate in a truly autonomous mode, management would either have to be recruited from sources other than the parent firms or, if spun off from the parents, it should be made clear that there is no continuing relationship or responsibility—short- or long-term. Hence, individuals whose skills are clearly marketable elsewhere should be sought.

One writer goes further by generalizing to the effect that "staffing decisions whereby owners send managers to work within their ventures will be complicated by an unwillingness to let go. Too many firms," it was concluded, "permit restive employees to use the 'revolving door' back to headquarters and the security of the owner's organizations. Because it is not possible for the best interests of the venture to be served by managers with unclearly focused loyalties and attention, the venture's managers may have to be recruited from outside. [Those in] key positions, in particular, must have unswerving loyalties to the well-being of the venture."[23]

This same scholar makes the point that insofar as a joint venture's board of directors (also the management group of a partnership?) is concerned, "the candidates ... are often managers operating so far down within the owner firms' hierarchies that they would never be considered for similar honors within their own organizations." Yet, the author goes on, "their skills and insights may be more compatible with the needs of the venture, especially in the early years of a venture's startup, than higher-ranking executives [who] ... are accustomed to dealing with the firm's external environments and the complexities of multi-business enterprise."[24] The study suggests a combination of rotation and continuity as different operating needs need attention as the venture develops its own capabilities. At the same time, some continuity in the partners' dealing with one another is required. "No number of codicils can overcome the benefit of prolonged exposure by the permanent members of the partners' delegations to each other. The longer they work together harmoniously, the less need they will feel for recourse to legal documents to establish a homogeneity of visions concerning their venture's purpose."[25]

One work on joint-venture management lays down several management guidelines for improving the chance of a joint venture's success. It would appear that a partnership or collaboration could likewise be included.

1. The more a venture can be run *as if* it has only one parent, the simpler will be the management task.[26]

2. There is a very strong [positive] relationship between the success of a joint venture and the autonomy of its general manager.[27]
3. The [joint] venture's board of directors ('a shared body') [should] be given responsibility for the strategic direction of the venture, but . . . one parent [should] dominate the venture's operating side."[28]

In fact, arguments can be made to the contrary in specific situations. For example, if both partners feel that they have important stakes in an enterprise and significant contributions to make, to manage it as though it had only one parent would surely doom the effort. The same could be said relative to the autonomy of the general manager, unless in an "independent" joint enterprise. Strategic direction should certainly be in the hands of the board of directors or, in the case of a partnership, a management group, on which both partners are represented, and it is not at all clear tthat one parent should dominate the operating side. Dominance of one partner over the other in operations depends upon the relative expertise of the partner firms in various areas and the relative cost at which the two firms are willing to contribute their expertise to the enterprise.

Perhaps of an importance equal to that of staffing is an explicit discussion and agreement at the start of an enterprise regarding such management processes as planning, financial reporting, the monitoring of performance, maintaining a management information system, product testing (i.e., quality control), and industrial relations. If one has a wide range of potential partners, it is possible, as suggested under success factor #1 discussed above, to select a partner with an overall organizational culture and managerial style compatible with one's own. If so, the working out of a general operational protocol or standard operating procedure for the joint enterprise should be relatively easy. Even in the case of dominated, divided, or independent joint enterprises, some degree of coherence in operational procedures between the joint enterprise and those of the partner firms is urged in order to avoid suspicion, mistrust, and conflict. Partners who have assets at risk in a joint undertaking need to be able, at least generally, to understand what is going on—that is, how their assets are being used.

Explicit Agreement as to Conditions of Termination

To emphasize a point made earlier, termination of a joint enterprise need not be a signal of failure. *Contractual* joint ventures always terminate at some point, either by mutual consent or upon completion of a given undertaking. And, at some point a purely contractual relationship of a more limited variety may be seen as preferable to *any* form of joint enterprise—contractual, equity, or partnership—by both partners.

In all cases, initial agreement as to what happens upon termination is reassuring to all parties. The argument that one does not normally discuss divorce when one marries is not a valid analogy. (In any event, possibly they should, and the wise pobably do!) If partner firms place valuable assets at risk within a joint undertaking, reasonable assurance that those assets—or what remains of them—will return to their control in event of termination reduces the level of unease. This observation is particularly true for such assets as rights under patents, trademarks (including the use of company names), copyrights, and trade secrets (i.e., unpatented proprietary know-how). Most firms now insist upon the return of such assets if the joint enterprise is restructured in such a manner as to become an independent competitor (or could become such) in one or more markets important to the partner. An example would be when the partner from whom those rights originated can no longer control their use to the degree felt to be essential, as in a trademark license. Indeed, some degree of control of protected rights may be legally required, as under U.K. law, where the licensor of a trademark must maintain some degree of quality control. By not capitalizing such rights in the equity of a joint venture, but rather licensing their use under separate contract and restricting their use by such means, may be far safer for the concerned partner. Hence, less risk is perceived than in a joint-venture undertaking. The general rule is that it is best to place those assets of greatest concern to the originating partner under contract and to avoid their capitalization in a joint venture.

Termination may be triggered by prior agreement if production, sales, profit, product quality, or some other *objective* measure fails to be fulfilled, or if the joint enterprise is managed contrary to agreement, particularly in respect to financial structure, transfer-pricing, non-disclosure of proprietary know-how, product mix, markets, distribution channels, availability of financial records, plant access by a partner's representatives, or failure to submit required reports to one or both partners.

Explicit Agreement for Arbitration

It is possibly true that many, if not most, conflicts between partners in a joint enterprise which cannot be resolved other than by outside intervention, such as arbitration, fail in the sense previously defined in this paper. Recrimination, ill will, and a complete break in relationship between firms is very likely to follow.

The choice of method for resolving conflict which cannot be resolved by internal discussion and negotiation (or friendly mediation) is between judicial decision in a court of law or by arbitration. My own preference, based on joint enterprises with which I am familiar, is arbitration. An international joint-venture enterprise involves at least two legal systems by definition. The law in the two countries may differ substantially, and the question always arises whether the courts in one country will enforce a foreign court decision against a local party. In any event, the uncertainty of the relevant law, delay, and cost tend to mitigate against dependence on judicial decision.

On the other hand, arbitration—depending upon the specific process referred to in the underlying agreement—can be relatively informal, fast, and cheap. This subject was dealt with in an earlier chapter.

Suffice to add here that even though neither partner feels that any conflict is likely to go as far as arbitration, the provision for objective and equitable resolution of conflict is reassuring. Insistence by a partner upon resolution of conflict in its parent country or before a local arbitrator can be very high-cost in terms of perceived risk by the other partner, particularly if the latter is not completely assured that the local courts or arbitration process are really insulated from political, nationalistic, or financial influence. One is often told by international executives that having a good, mutually acceptable system in place for resolving conflict greatly enhances the probability that such conflict will never arise.

Establishment of a System for Facilitating Separation of Costs

Several scholars have noted that by establishing a facility completely separate physically from either partner, the possibility of ascertaining objectively determined costs and revenues is improved considerably, and, hence, the opportunity for conflict reduced. If a joint enterprise, particularly of the equity or contract-based joint-venture type, has assets located physically within the facility of one partner, there is the problem of allocating to the activities of the venture a percentage of the costs shared with other production taking place within the same facility. Building, equipment, maintenance, supplies, and even supervision and labor, may not be used or assigned on a 100 percent basis to the joint-enterprise activity. Hence, division of costs may be arbitrary. If the joint enterprise were vertically integrated, on site, with the functions of one of the partners, independent and objective calculation of the value-added by the joint enterprise itself may be rendered substantially more difficult and also, therefore, the revenue generated thereby. In such cases, the other partner may feel compelled to monitor output very closely in order to be assured that tallies are correct in that no independently generated payroll, shipping lists, or invoices would necessarily be available. Such considerations may seem marginal, but possibly worth thought if reasonable options exist. This problem possibly does not exist for the typical partnership or collaborative agreement in that there are no production facilities apart from those of the partners, each of whom perform designated functions in their own facilities and, in so doing, maintain cost centers.

TECHNOLOGY TRANSFER, STRUCTURE, AND CONTROL

In the final analysis, the international transfer of technology cannot be divorced from the structure within which it moves. The choices are really five-fold: (1) the wholly owned or majority-owned subsidiary (the foreign direct investment package), (2) the equity joint venture, (3) the contractual joint venture, (4) the partnership or strategic alliance, and (5) the pure contract (license, technical assistance, etc.). The distinctive feature of the latter is that it is limited to a specific project, technology transfer, or use of a proprietary right over a stipulated time. An integral part of any discussion of these alternative transfer modes is the degree of control an interested party may exercise over the use of the transferred technology or right. It is my fond hope that the argument that effective control cannot be equated with degrees of ownership in the international case has been persuasive.

NOTES

1. Wolfgang G. Friedman and George Kalmanoff, eds., *Joint International Business Ventures* (New York: Columbia University Press, 1961), pp. 155–56.
2. *Ibid.*, p. 167.
3. James W. C. Tomlinson, "A Model of the Joint Venture Process in International Business," unpublished Ph.D. dissertation submitted to the Sloan School of Management, Massachusetts Institute of Technology, 1969, p. 265.
4. Peter Killing, *Strategies for Joint Venture Success* (London: Creom Helm, 1983), pp. 13, 18.
5. *Ibid.*, p. 55.
6. *Ibid.*, p. 83.
7. Daric Iacuelli, "Management Factors in a Partnership's Joint Venture: Italy, A Case Study," unpublished Ph.D. dissertation submitted to the Sloan School of Management, Massachusetts Institute of Technology, 1970.
8. See Richard D. Robinson, *The Internationalization of Business* (Hinsdale, IL: Dryden Press, 1984), pp. 140–46, for further discussion of this subject.
9. Killing, *op. cit.*, p. 56.
10. *Ibid.*, p. 56.
11. Iacuelli, *op. cit.*, pp. 230–31.
12. Kathryn Rudie Harrigan, *Managing for Joint Venture Success* (Lexington, MA: Lexington Books, 1986), p. 80.
13. And, I would add, shared learning, as well as economies of scope derived from the accumulation of experience, know-how, and assets over time (e.g., market acceptability by reason of a well-established name which can be used to facilitate market entry for a variety of products).
14. Harrigan, *op. cit.*, p. 30.
15. Iacuelli, *op. cit.*, p. 234.
16. Iacuelli, *op. cit.*, pp. 232–33.
17. Harrigan, *op. cit.*, p. 71.

18. Published originally in Richard D. Robinson, "Ownership Across National Frontiers," Sloan School of Management, Massachsetts Institute of Technology, Working Paper #368-69, February 1969. A somewhat different version is suggested by Harrigan, *op. cit.*, pp. 29-52.
19. Killing, *op. cit.*, p. 21.
20. *Ibid.*, p. 63.
21. Harrigan, *op. cit.*, p. 27.
22. Killing, *op. cit.*, p. 48.
23. Harrigan, *op. cit.*, p. 79.
24. *Ibid.*, pp. 78-79.
25. *Ibid.*, p. 79.
26. Killing, *op. cit.*, p. 15.
27. *Ibid.*, p. 22.
28. *Ibid.*, p. 54.

A Last Word

In the preceding pages we have explored many of the intricacies and perplexities involved when technology flows across national borders. Much of the problem arises because of our inability to assign an unambiguous value to a unit of transferred technology. As discussed early on, for that technology transferred internally by a firm from one affiliated enterprise to another, there is no market determined value. It is largely arbitrary. Even for much of the technology moving externally at a negotiated price between unrelated enterprises, the market is very imperfect. Absence of true competition among the suppliers of technology, limited knowledge of technological alternatives by both technology supplier and recipient, and the intervention of governments all contribute to the imperfect nature of the international technology market.

A related factor is the increasing awareness by people everywhere of the importance of externalities (effects outside of the supplying or using enterprises) generated by transferred technology. These externalities include such phenomena as employment displacement, income redistribution, shifts in military and political power, and redefinition of values and world views. These factors can be viewed as positive or negative, depending upon how one views the sweep of human history and differentiates good from bad in that evolutionary process. There is also the risk of environmental degradation and injury to the health of employees, consumers, and the general public. The market cannot evaluate these externalities, although the political sector may attempt to do so, however imperfectly.

Technology developed in one society and implanted in another without adaptation may be likened to transplanting a living organ from one body to another without testing compatibility. If the transplanted organ proves to be incompatible, it is very likely to be rejected, thereby bringing injury, even death, to its host. Many examples of injury sustained by the international transfer of unmodified technology can be cited.[1]

- The implant of modern medicine to save and prolong life where the economic wherewithal to sustain life is absent;
- The transfer of devices rendering the control of a population by an authoritarian and repressive minority more intrusive and pernicious;
- The sale of modern weapons to a country without the capability or will to control their use for the public good;
- The transfer of labor-saving technology to a society suffering from under or unemployment;
- The marketing of a product requiring an educated literate public, if it is to be used without ill effect, where such a public is largely nonexistent;
- The setting up of a manufacturing plant using potentially dangerous processes or products in a society lacking adequate inspection and enforcement capabilities; and
- The sale to a relatively poor country of technology that diverts local resources from more to less essential uses, as defined by the priority of local needs perceived by the recipient society.

It is very easy to condemn much of modern technology as demeaning to the individual and destructive of the environment in ways that are possibly irreversible. Much of modern technology, some argue, has been driven by the desire to achieve maximum economies of scale and ever higher levels of consumption no matter what the consequences to society or nonrenewable resources. The result has been large unwieldy corporate "tribes," densely crowded megacities, and unrestrained plunder of underpriced irreplaceable resources. The outcome can be the displacement, overuse, and pollution of the natural environment and the collapse of meaningful individuality, as society is bludgeoned into acceptance of these conditions through ubiquitous commercial advertising and political rhetoric or worse, cohersion. The world becomes more homogenous. The processes and products of the new technology are too complex for producer or consumer to understand or evaluate. The result? A web of increasingly pervasive law and regulation. One must be protected from unwitting self-imposed injury. The individual's space shrinks, psychologically and socially—and possibly politically as well. Meanwhile, the technology gnaws away at the human heritage of pure air and water, abundant natural resources, and untrammelled space, and in their place deposits toxic effluence and intensified misery.[2]

One can easily conclude that the process is terminal. Even though the commercialization of really significantly new technology may have slowed for a time, as I have argued, the accelerating dissemination of existing technology worldwide is a fact. It is probably true that the enthusiastic acceptance of that technology by those outside the societies to which it is indigenous has, at times, done much damage. However, there is also evidence of increasing worldwide awareness of the danger inherent in the unfettered transborder flow of technology. We are witness to an ongoing effort to suggest codes and guidelines that national governments or regional authorities can apply. The world has become alert and is beginning to act. One can be mildly optimistic. It would appear that internationally transferred technology is being monitored ever more carefully by both the transferring and receiving societies in a positive manner. Economic efficiency and maximum consumption are no longer the sole tests, or even the major ones. Therefore, we can speak in good conscience about ways of facilitating the international movement of technology.

The very idea of economy of scale is being redefined. Large size need not be equated with least cost. Nor is the most complex technology considered necessarily the best. The notion of "intermediate technology" has become commonplace.[3] The miniplant and remanufacturing are well-known concepts. Ideas of corporate liability for both environmental and human injury are causing many to rethink what technology should be transferred internationally, what transfer devices (organizationally speaking) should be used, and how control to prevent the misuse of transferred technology can be enforced.

Meanwhile, there is much talk of ways to reduce the cost of international technology transfer, particularly from the relatively affluent

216 / A LAST WORD

to the relatively poor societies. A variety of cost-reducing measures suggest themselves.

- The creation of large technology data banks encompassing the world's technology and providing information as to who has the capability of doing what and under what conditions in access to the relevant technology possible. Improved knowledge on the part of prospective technology buyers would make the international technology market more competitive. Access to such data banks should be made possible for LDC governments and firms at nominal cost. Moves in this direction have been made by the United Nations, the European Community, the Andean Common Market, and various private commercial firms. The largest technology data banks are possibly those maintained by the major Japanese general trading companies, but cost of access is hardly nominal.

- The creation of publicly and privately funded laboratories to study the technical alternatives in respect to new technology appearing in the market. Such alternatives would include the possibility of using different skill and labor/capital mixes, employing different materials, reducing the scale of production, downscaling product size, lessening the energy input (or altering its kind), altering performance specifications, and modifying designs so as to be more appropriate for different markets (particularly those of the poorest countries) and minimize the possibility of injurious misuse and environmental damage. The full range of alternative technologies might thus be revealed and the market enhanced.

- Restricting the use of patents and trademarks in LDC countries so as to protect only that technology actually being used locally (after a certain grace period), whether via license, technology transfer contract, or in association with direct foreign investment.

- Inducing governments of major technology countries to reduce or waive completely domestic taxes levied on royalties and fees arising from technology transferred to LDCs and on the salaries of their citizens whose principal activity is technology transfer to LDCs.

- Enforcement by LDC governments of awards rendered by foreign arbitration tribunals when such arbitration is conducted according to agreement between the disputing parties, and if one of them be a government or state-owned enterprise, the waiving of any claim to sovereign immunity which might otherwise bar execution of such an award.

- Establishing an international agency (possibly within the World Bank family) which would register corporations that commit themselves to adhere to specific guidelines relating to the international transfer of technology and which maintain a record of performing in accord to those guidelines. Such "registered corporations" would be exempt from any performance bond or bank guarantee requirement and from the practice of withholding payments to assure performance under a technology transfer contract.
- A legally binding international understanding among countries (or bilaterally) that when a host government approves the importation of a given technology and undertakes periodic inspection of facilities using that technology to assure itself that all reasonable precautions are being undertaken to prevent injury, the foreign supplier of that technology will be shielded from any subsequent liability. (Perhaps this facility should be open only to "registered corporations.")
- Development of a scheme whereby the flow of students and scholars and technicians into the technical and scientific institutions of the major technology exporting countries is facilitated, plus refusal by the host countries to grant the visitors residence and/or work permits for a given number of years (say twenty) once their training has been completed.
- An agreement among the major technology exporting countries that residence and/or work permits shall not be given to "highly skilled" persons under a specific age (say forty) arriving from LDCs, unless the full cost of the education and training received at public expense in their home country be reimbursed to the appropriate public authorities.

Upon reflection, one will note that these suggestions would all have the impact of reducing technology transfer cost, enhancing the benefits of a transfer to technology suppliers (via reduced tax, transfer cost, and risk), or providing greater assurance to the technology user that what is received is more likely to be appropriate for a particular use and less injurious than might otherwise be the case.

It will require these sorts of collaborative efforts if the international transfer of technology is to result in the general uplift of the human condition. The architects of such efforts must, however, understand well what is involved in the transfer process. Otherwise, unintended and undesired results could flow from the effort. The purpose in the writing of this book is thereby revealed.

NOTES

1. Explored in considerable detail by Denis Goulet, *The Uncertain Promise, Value Conflict in Technology Transfer* (New York: IDOC/North America, 1977).
2. An eloquent statement of this general view is found in E.F. Schumacher, *Small Is Beautiful, Economics as if People Mattered* (London: Blond & Briggs Ltd, 1973). The argument is carried forward by Donella H. Meadows et al. in *The Limits to Growth* (New York: New American Library, 1972).
3. One need only refer to the publications of Intermediate Technology Publications, Ltd (London) and to the "Development and Transfer of Technology Series" of the UN Industrial Development Organization (UNIDO).

Index

A&E (architectural and engineering contract), 5, 162
Africa, South, 84
African Development Bank, 151
Agreements, technology transfer, 40, 74–76, 157–59; cost-plus-type, 159; fixed-price, 157–59; repricing, 158
Agrochemicals, ecologically dangerous, 88
AID program, U.S., 61
Airbus GIE, 173
Aircraft industry, 172, 177
ANCOM (Andean Common Market), 86, 216
Antitrust law, 85, 143
Arbitration process, international, 78
Asian Development Bank, 151
AT&T-Olivetti relationship, 176–77
Australia, foreign R&D units in, 122
Austria, 139
Automotive industry, 31; and remanufacturing, 69–70

Balance-of-payments effect, 63–64
Banks, development, 61
Battelle Memorial Institute, 43
Behrman, Jack N., 128–29
Belgium: foreign R&D units in, 120–21, 122; patents issued by, 133

Berne Convention of 1886, 142; signatory nations of the, 137
Boeing (company), 178
Brain drain issue, the, 26–27
Brazil, 125, 137, 143
British Commonwealth countries, 134, 135
Business, private, 23–24; corporate structure of, and the location of R&D units, 126–27; and the cost of technology transfer, 41–44, 62; and the reception of foreign technology, 73–80
—decision-making process within, 37–57; and the contract option, 38, 48; and external vs. internal transfer, 44–47, 56–57, 78; and government intervention, 37, 47–48
Business International, 112, 126
"Buy American" law, 157

Caisse Centrale de Coopération Economique, 151
Canada: foreign R&D units in, 120, 121, 123–24, 125; and the patent system, 145; protection of trade secrets in, 138; subsidies for R&D in, 113
Capital, developmental, 49

Capital-intensity, and perceived cost, 53. *See also* Technology, capital-intensive
Center for Policy Alternatives, 69
CFM International, 177
Chemical industry, 88, 89–90, 114
China, People's Republic of, 26, 50, 144, 145
Chuan, Tan Beng, 151
CIDC (Community Industrial Development Contract), 114–15
Coca-Cola (company), 138
COCOM (Coordinating Committee), 84
Collaborative agreements, 170. *See also* Joint enterprises; Joint ventures
Columbo Plan, 61
COMECON (Council for Mutual Economic Assistance), 42
COMECON countries, 51
Commonwealth Development Fund, 151–52
Community Patent Convention, 139–40
Community Patent system, 140
Concessionary prices, 102–3
Consortium, 170. *See also* Joint enterprises
Consumer-rights guidelines, 88
Contractor, Farok J., 51
Contracts, technology transfer, 29, 31, 48, 98, 160–62; arbitration clauses in, 78; vs. control via ownership, 48, 165–66; under the ICB process, 160–62; pure, 209–10; R&D as the subject of, 50, 113; technical collaboration, 96; turnkey, 6, 163, 175; turnkey-plus, 6, 163; and value-added chain analysis, 48–52. *See also* Agreements
Control: areas of, 196–97; mechanisms, 198–99; and ownership, 48, 165–66, 183–84; positive and negative, 195; and success, in technology transfer, 183–210
Copyright Act (1909), 137
Copyright protection, 132, 136–37, 142, 147. *See also* Protection, of internationally transferred technology
Corning Glass Works (company), 119–20

Corporate technology units, 117–23
Counterfeit goods, 142–44
CPC International, 120, 122
Cuba, 144

DATAR (Délegation à l'Aménagement du Territoroire et à l'Action Régionales), 113
Davidson, William H., 44, 45
Discovery diplomas, 144
Distributional effect, 65
Diversification, geographical, 190
Dollar, depreciation of the, 110
DuPont (company), 12, 124, 127

Eastern Europe, 50, 71, 144
Eastman Kodak (company), 122
EC member states, 114–15, 216; and international protection systems, 139, 140
EC patent, the, 115
Economic Commission for Europe, 71
Economic Development Fund, 62
Economic growth: long-term, 64–65; maximum, vs. maximum employment, 66
Economic theory, international, 24–25
Ecosystems, 67
ECU (European Currency Unit), 195
Education, 67
EEC (European Economic Community), 114
EEIG (European Economic Interest Grouping), 174–75
Egypt, 143
Electrical certification system, 71–72
Electrical products, European trade in, 71–72
Electronic: firms, 114, 176; technology, 119
Emigration policies, restrictive, 25–27
Endangered species, trade in products involving, 88
Entrepreneurial services, transfer of, 50
Environmental impact assessment requirements, 88
Equity: external transfers via, 165; foreign held, 86, 87, 174; purchase of, decline in the, 175–76. *See also* Joint enterprises

European Communities, the, 62, 71, 85, 88
European Community patent, 139–40
European Free Trade Area, 71
European Investment Bank, 114
European Patent Convention, 139, 140
Exchange rates, fluctuations in, 161, 172, 188
Exclusivity provisions, 75
Export controls, 37, 75, 80
Export-Import Bank, 162
Exxon Chemical, 120
Exxon Corporation, 119, 120

FDI (foreign direct investment), 31, 44, 52–53, 166, 209; vs. external technology transfer, 47; in LDCs, 87–88
Financial services, 49–51
Fischer, William A., 128–29
Flexible pricing, 158
Ford Foundation, 26
Foreign Credit Insurance Association, 162
France: GIE in, 173; patents issued by, 133; protection of trade secrets in, 138; R&D units in, 113, 125
Friedman, Wolfgang G., 169–70, 172

GATT (General Agreement on Tariffs and Trade), 72, 98; Uruguay Round of the, 142
Germany, West, 68, 133
GE-SNECMA collaboration, 176–77
GIE (Groupement d'Interest Economique), 173
Global technology units, 117, 119, 121–22, 123, 127
Government: influence of, on R&D projects, 115–16; motivations of, to transfer technology, 23–24; multi–, organizations, intervention by, 88–91;
—owned enterprises, on the demand side, 61, 62, 74;
—s, host, and export restrictions, 75; subsidies, 92–105, 113. *See also* Intervention, government, in technology transfer
Growth-generating effect, 64–65

Hard goods, 50. *See also* Technology, embodied

IBM (company), 119, 121
ICB (international competitive bidding), 151–63, 178; and arguments for sourcing from abroad, 152; the bidding process of, 153–56; negotiation and, 156; and the pricing problem, 157–62; reasons for, 153
IEC (International Electrotechnical Commission), 72
Import duties, on capital equipment, 102
India, 125; Union Carbide in, 89–90, 124
Indigenization, 32
Indigenous technology units, 117, 119–121, 122–23, 126, 127
Indonesia, 86, 143
Industrialization Fund for Developing Countries, 152
Information, gathering and sorting, 49
Initiating mechanisms, 115, 116
Innovation, process vs. product, 45
Innovational effect, 65
Innovation process, government influence on the, 115–16
Institute for Intermediate Technology, 61
Inter-American Development Bank, 151
International Bank for Reconstruction and Development, 151
International Chamber of Commerce, 78
International Development Association, 151
International Finance Corporation, 151
International Patent Institute, 139, 140
Intervention, government, in technology transfer, 28, 37, 47–48, 76–77, 83–105; via foreign direct investment, 56; and foreign R&D units, 126, 127; four levers of, 91–105; and the international bidding process, 83; and labor-saving technology, 63; and multigovernment interventions, 88–91; and the protection of proprietary rights, 83; regulation as

a device of, 62, 91–105; and subsidies, 91–105; and tax incentives, 92–105; uncertainty reduction as a device of, 93–105
Inventor's certificate, 144–45
Iraq, 169
ISO (International Organization for Standardization), 72
Italy, patents issued by, 133

Japan, 114, 115, 117; foreign R&D units in, 121, 122, 123, 125; Ministry of International Trade and Industry, 141; and protection of trade secrets in, 138; and technology protection in LDCs, 147
Japan Aircraft Development Corporation, 178
Japan Commercial Airplane Company, 178
Japan Economic Journal, 176
Johnson & Johnson (company), 119–20, 121, 225
Joint enterprises, 85, 165–80, 183–210; agreements for arbitration in, 208; conditions of termination in, 207–8; control and success in, 183–210. *See also* Control; costs and benefits of, 200–203; decision-making in, 191; definition of, as a profit center, 194; equity, 96, 167–70; goal incongruities in, 192–93, 200; loyalty in, 204–6; management in, 189–92, 204–6; "organizational culture" of, 190–91; separation of costs in, system for, 209; setting of transfer prices in, 194–95; and staffing, 204–6; strategic alliances, 39–40, 170–80, 209; and technical collaboration agreements, 167, 168, 183
Joint ventures: contractual, 96, 165, 167–70, 207; divided, 183; dominated, 183; equity, 209; independent, 183, 188; shared, 183

Korea, 143

Labor, opposition to new technology, 28
Labor-intensity: analysis of, at the individual firm level, 25; and appropriate technology, 67; def., 11; and maturity, 11–12; and perceived cost, 53
Latin American countries, patent protection in, 134
Law, international, 78, 90, 93
LDCs (less developed countries), 29, 88; expenditures for imported technology, 113; governments of, intervention by, 85–88, 216–17; multinational corporations based in, 34; patent protection in, 135; R&D located in, 112–13, 124–25, 127; technology protection in, 145–48; technology transfer out of, 24, 27; technology transfer to, 51, 74. *See also* NICs
Liability, product and waste, 88–90
LIBOR (London Inter-Bank Offer Rate), 194
Liechtenstein, 139
Luxemburg, 133

McFetridge, Donald G., 44, 45
Madrid Agreement of 1891, 141, 149
Malaysia, 143
Management, and joint-enterprise success, 189–90, 204–6
Market: development, 48, 49; testing, 48, 49, 50
Martin Marietta (company), 178
MDDT Corporation, 178
Media, public, 6
Mexico, laws regarding patents in, 134, 147
Minebea Company, 176
Ministry of International Trade and Industry (Japan), 141
Modernization, 68
Monaco, 139
Monopolization, of markets, 94
Monopoly "rent," 38, 44, 47
MRLS-3 (Multiple Launch Rocket System), 178
Multinational corporations, 31, 46–47; and the contract option, 47–48; erosion of control by, 29

National Semiconductor (company), 176
NATO (North Atlantic Treaty Organization), 178–79
Natural law, 14

Natural resources, 63, 67, 116, 215
Nestlé Corporation, boycott against, 88–89
Netherlands Finance Company for Developing Countries, 152
NICs (newly industrialized countries), 31; improvement technology in, development of, 32; and modes of technology transfer, 166; and patent protection, 28. *See also* LDCs
Nigeria, 68, 143
NMB Semiconductor Company, 176
North-South dialogue, the, 146
Norway, 139
NPOs (non-profit organizations), 23–24

OECD (Organization for Economic Corporation and Development), 31
Otis Elevator, 119
Overseas Economic Cooperation Fund, 152

Paper machinery industry, 70
Paris Convention for the International Protection of Property Rights (1883), 139, 141, 146, 148; 1967 amendment to the (the Stockholm Convention), 144–45
Partnerships, 165, 167. *See also* Joint enterprises
Patent: ownership, in joint enterprises, 198, 207; policy, 38–39, 85; protection, 132–35, 138–44. *See also* Protection, of internationally transferred technology:
—s, confirmation, 134–35; systems, 28, 128, 140
Patent Cooperation Treaty (1975), 140, 141
Peace Corps, the, 26, 61
Pharmaceutical firms, 114, 172
Philippines, the, 143
Price adjustment, main methods of, 160
Pricing, flexible and retroactive, 158
Private Overseas Private Investment Corporation, 162
Procurement: alternative methods of, 153; negotiated, 153; single-source, 153

Production, 48, 49; batch, 15; preliminary, 48, 49, 50
Product life cycle theory, 34
Protection, of internationally transferred technology, 131–49, 216; in centrally planned economies, 144–45; four types of, 132–38; international systems of, 138–42; in LDCs, 145–48; management problems arising from, 148–49; and the Paris Convention for the International Protection of Property Rights (1883), 139, 141, 144–45, 146, 148; and trade in counterfeit goods, 142-44
Public revenue effect, 64

R&D (research and development), 5, 43, 78, 138, 172; as capital-intensive, 25; and the commercialization of new technology, 27–28, 32; cost of, 27, 109, 116, 123–24, 127; decentralized forms of, criticism of, 109; government influence on, 113–14, 115–16, 125–26, 127; in high-technology companies, 111; in LDC operations, 112–13; and marketing strategies, 34; military, 32, 107; national expenditures for, 107–8; siting of, in foreign countries, 107–28; subsidies for, 113–14; and tax legislation, U.S., 55–56; and transfer costs, 7; and transfers under contract, 50; units, foreign, 117–23; and value-added chain analysis, 48, 49
Regulation, and intervening governments, 62, 91–105
Remanufacturing industry, 69
Research Development Corporation, 114
Restrictions: application, 75; export, 37, 75, 80; field-of-use, 75
Restructuring mechanisms, 115, 116
Retroactive pricing, 158
Rice Institute, 61
Rome Treaty, the, 140
Ronstadt, Robert, 128
Root, Franklin R., 51

Safety codes, industrial, 88
Saudi Arabia, 87
SDR (Special Drawing Rights), 195

Singapore, 25–26, 143
Societé Belge d'Investissement International, 152
Soft goods, 5
Soviet Union: R&D facilities in, 113; technology protection in, 144, 145
Standards Code (Agreement of Technical Barriers to Trade), 71
Steel production, 31
Stockholm Convention, the, 144–45
Strategic alliances, 39–40, 170–80, 209. *See also* Joint enterprises
Subsidies, government, 92–105, 113
Sustaining mechanisms, 115, 116
Sweden, 139
Swedfund, 152
Switzerland, 139
Systems Dynamics Group (Massachusetts Institute of Technology), 29

Taiwan, 143
Taxation, 64, 87, 98; and income from technology transfers, 84; and the location of R&D, 55–56
Tax Reform Act (1986), 56
Technological innovation, commercialization of, 27–32
Technology: adaptive, 33; applications, 114; appropriate, 66–67; associated, 68; basic, 12–13; branching, 12–13; bundled, 38; capital-intensive, 42, 65, 66, 74; centrality of, 14; completeness, 17, 57; complexity of, 14; continuity of production of, 15; core, 4, 14; data banks, creation of, 216; design, 17, 18; dimensions of, 11–19; disembodied, 5; dynamism of, 12, 57; electronic, 119; embodied, 50, 69; environmental specificity of, 13, 57; factor substitutability of, 13, 57; firm specificity of, 15–16; grant-back, 75; improvement, 32; incremental, 12–13; intermediate, 67, 215; labor-intensive, 42, 65, 67, 69; licensing of, 68; major improvement, 12–13; maturity of, and labor-intensity, 11–12, 27, 45; non-proprietary, 5; peripheral, 4, 14, 44; pharmaceutical, 64; primacy of, 17, 57; process, 15, 17, 18; product, 15, 17, 18; product adaptive technology, 18; proprietary, 5; relative importance of, 12–13; scale specificity of, 14, 57; state of the art, 67; susceptibility of, to reverse engineering, 15, 57; transformation, 33; unbundled, 38, 162, 166; user technology, 17, 18, 65
—transferred, protection of, 131–49; four types of, 132–38; international systems of, 138–42; management problems arising from, 148–49
Technology transfer, international: agreements, 40, 74–76, 157–59; "bargaining window" in, 8–9; code of conduct for, 90–91; contracts. *See* Contracts, technology transfer; external impacts of, 62–66; financing for, 76, 77–88; general definition of, 10; indirect, 6; internal and external, 37, 38–48, 165; international standards for, 70–73; modes of, 165–80; negative impacts of, 213–15; packages, 3–6; policy options, 17–19; political impact of, 65–66; social impact of, 61; units, 117–18, 120
—cost of, 6–10, 41–44, 215–17; on the demand side, 67–77; and firm specificity, 16; internal vs. external, 76; perceived, 42, 53, 74; perceived and actual, 42; and the supplying firm, 41
—the demand side of, 61–80; and the absorptive capacity of the receiver, 67–77; and appropriate technology, 66–77; and external impacts of the transfer, 62, 63–66; and government intervention, 63, 75, 76–77; and restrictions in transfer agreements, 74–76
—the supply side of, 23–56; and the brain drain issue, 26–27; and the commercialization of technological innovation, 27–32; direction of flow in, 24–26; and internal vs. external transfer, 38–48; model of, 53–57; sources of, 23–24; and technology flow from LDCs, 32–34; and value-added chain analysis, 48–52
Thailand, 143
Tied-buying provisions, 74–75
Tokyo Round of Multilateral Trade Negotiations, 71–72

Trade: agreements, bilateral, 98; controls, and multigovernment intervention, 88; free, areas of, 98; in products involving endangered species, 88; sanctions, 84; secrets, 132, 137–38; and U.S. foreign policy, 85
Trademark Registration Treaty, 141
Trademarks, 132, 135–36, 141, 198; and technology protection in LDCs, 147, 216. *See also* Protection, of internationally transferred technology

UCC (Universal Copyright Convention), 142
Uncertainty reduction, 93–96
Unilever (company), 125
Union Carbide (company): —Bhopal tragedy, 89–90, 172; New Delhi laboratory of, 124; R&D units of, 120, 122; technology transfer units of, 119
United Kingdom, 120, 141, 207
United Nations, 23, 61, 72, 216; adoption of the "Declaration on the Establishment of a New Economic Order," 90; and the brain drain issue, 26; Commission on Trade Law, 78; efforts to create an international safety code, 90. World Health Organization (WHO), 89
United States: balance-of-payments deficit, 172; Copyright Act of 1976, 136–37; copyright protection in, 136–37, 143; foreign policy, 85; loss of technological and manufacturing dominance, 172, 176; patent protection in, 134–35, 141, 143; protection against counterfeiting in, 143; protection of trade secrets in, 138; trademarks issued in, 135–36
Universal Copyright Convention, 137, 145

Value-added chain analysis, 48–52
Value-added effect, 63, 64

Wastes, hazardous, 88, 94
Wheat Institute, 61
WHO (United Nations World Health Organization), 89
Woo, Hing Kwok L., 129
World Bank, the, 61, 157, 159–60
World Intellectual Property Organization in Geneva, 140, 141

Yugoslavia, 86

About the Author

Richard D. Robinson is the George F. Jewett Distinguished Professor of Business at the University of Puget Sound as well as professor emeritus of management at MIT's Sloan School, where he founded the international management concentration. He has been a member of the Massachusetts International Trade Council, National Committee on U.S.-China Relations, and served on the Board of Directors and Executive Committee of the International Business Center of New England. He has also served as president and dean of the fellows of the Academy of International Business, a member of the National Advisory Counsel of *Business*, a member of the Board of Editors of the *Journal of Economics and Business*, and a member of the Editorial Board of *International Marketing Review.* Dr. Robinson has published extensively.